Amorphous Chalcogenides

Amorphous Chalcogenides
The Past, Present, and Future

Victor I. Mikla, Victor V. Mikla

Institute for Solid State Physics & Chemistry
Uzhgorod National University
Uzhgorod
Ukraine

ELSEVIER

AMSTERDAM • BOSTON • HEIDELBERG • LONDON • NEW YORK • OXFORD
PARIS • SAN DIEGO • SAN FRANCISCO • SINGAPORE • SYDNEY • TOKYO

Elsevier
32 Jamestown Road London NW1 7BY
225 Wyman Street, Waltham, MA 02451, USA

First edition 2012

British Library Cataloguing-in-Publication Data
A catalogue record for this book is available from the British Library

Library of Congress Cataloging-in-Publication Data
A catalog record for this book is available from the Library of Congress

ISBN: 978-0-323-16539-6

This book has been manufactured using Print On Demand technology. Each copy is produced to order and is limited to black ink. The online version of this book will show color figures where appropriate.

Transferred to Digital Printing in 2016

Dedication

To the memory of my mother, my father, and my brother

Contents

Preface

Historically, selenium is one of the oldest and best-studied semiconductor materials, and it has a number of unique and practically useful physical properties, such as photoconductivity. It has served the photocopying industry for over three decades, and it has no analogs among other solid state materials in the purely academic sense. It has a wide spectrum of various applications and has been used successfully in xerography, photocells, photorectifiers, etc.

The first automated commercial office copier marketed in 1959 used amorphous selenium photoreceptors and revolutionized document reproduction. Xerox soon became a multibillion dollar company.

Less known is the use of amorphous selenium photoreceptors in X-ray imaging in a process called xeroradiography: it is the photocopying of a body part using X-rays. The X-ray photoconductivity of amorphous selenium was discovered during the early selenium development work at the Battelle Memorial Institute, a nonprofit research organization in Columbus, Ohio, in the 1940s. Xerox became involved in medical imaging by introducing a commercial xeroradiographic system for medical imaging in the early 1970s. Xeroradiography became obsolete by the mid-1970s. Somewhat later, scanned electrometer readouts were used in commercial digital chest X-ray imaging systems to enable the digitalization of the X-ray image. The readout technique significantly improved and modernized xeroradiography. At the same time, the fundamental xeroradiographic principle remained unchanged: the photoreceptor surface was first charged, like in the xerographic process, then it was selectively photodischarged by the incident X-rays passing through the object. The charge distribution was then a suitable readout. The true modernization and transformation of xeroradiography occurred in the form of a digital flat-panel X-ray image detector. The availability, usefulness, and convenience of such a readout technique inevitably lead to development of amorphous selenium-based flat-panel detectors. This system does not rely on the xerographic principle that involves charging and photodischarging the photoreceptor and reading the remaining charges. Since they are flat, one of the attractive advantages of the flat-panel X-ray detectors is that they serve as convenient direct replacements for the film cassettes used at present.

Amorphous selenium, like other chalcogenide glasses, is p-type: holes are more mobile than electrons. However, amorphous selenium has two important attributes that make it an exceptional case within the class of chalcogenides: first, both holes and electrons can drift in this material and both contribute to the photoconductivity. Second, the electrical properties of amorphous selenium are particularly sensitive to small amounts of impurities.

There was also interest in developing selenium-based photoconductive targets for TV pickup tubes. X-ray photoconductivity was recognized as an important attribute during the 1960s and 1970s, which led to the commercialization of amorphous selenium X-ray medical imaging systems. In the following paragraphs, we highlight some of the basic reasons for amorphous selenium's success as an X-ray photoconductor.

First, a reasonably thick layer is able to absorb X-rays and generate charge carriers, and amorphous selenium exhibits good X-ray photoconductivity.

Second, amorphous selenium can be readily coated by conventional vacuum deposition over large areas with good uniformity: X-ray image detectors need to be larger than the body part to be imaged since X-rays cannot practically be focused.

Third, both holes and electrons are mobile. This is a distinct advantage because X-rays are absorbed throughout the bulk of the layer.

Fourth, unlike many other amorphous solids, charge transport over the time scale of interest at room temperature is nondispersive for both electrons and holes. Researchers and engineers have been able to model and predict the behavior of selenium-based devices by simply using shallow and deep sets of traps. In the late 1980s in Japan, Kenkichi Tanioka et al. developed a practical amorphous selenium photoconductive target, called a high avalanche rushing photoconductor (HARP), which they eventually used in commercial TV pickup tubes. There is much current interest in using avalanche multiplication for amorphous selenium in all solid state photoconductive structures with electronic readout for various imaging applications, including applications in medical imaging. However, the overall use of the selenium material itself in these technologies is unlikely to reach the level that it did during heyday of xerography.

My interest in the physics of amorphous solids developed over a period of years within the stimulating environment at Uzhgorod University, under the guidance of Prof. Dmitrij Chepur, Prof. Vladimir Lendyel, and Prof. Vladimir Slivka. The Laboratory of Optical Data Storage has produced many important contributions to this field. I am indebted to many of my colleagues at the Institute for Solid State Physics and Chemistry, but I am especially grateful to Prof. Alexander Kökineshi, supervisor of my first (PhD) dissertation, whom I have had the pleasure of working with on diverse aspects of research on amorphous solids. Additionally, I wish to thank Prof. Safa Kasap, Prof. Hellmut Fritzsche, Prof. Mihai Popescu, Prof. Keiji Tanaka, Prof. Kenkichi Tanioka, and Prof. Marty Abkowitz for many stimulating discussions over the years.

Finally, I am deeply indebted to Dr. Lisa Tickner (Publishing Director at Elsevier) and Dr. Donna De Weerd-Wilson (Head of S&T Books at Elsevier). Both contributed their knowledge, competence, and interest to this edition. I am grateful to Dr. Lisa Tickner for encouraging me to expand the scope of this text to bring it up-to-date with a field of science that is so rapidly advancing. On a personal level, I found Dr. Lisa Tickner and Dr. Donna De Weerd-Wilson to possess the helpful and pleasant personalities that people with their intellect so often exhibit.

I would like to express my sincere thanks to Priya Kumaraguruparan and Mr. Paul Prasad (Project Managers, Elsevier) for their patience, continued interest, and helpful comments, which have made it possible for this book to reach completion.

However, I wish to express my deep appreciation to my wife, Ottilia, whose indispensable support made this book possible.

Victor I. Mikla
Uzhgorod
February 2011

Introduction

The research efforts of physical scientists over the past 60 years have helped secure a promising commercial future for amorphous chalcogenide semiconductors, in areas as diverse as X-ray image formation of the human body, high-definition TV pickup tubes, transparent element imaging in the infrared region of spectrum, submicron optics, and many others.

More than a century ago, in 1899, it was suggested by Charles Duell, head of the US Patent Office, that the patent system should be shut down because everything that could be invented already had been. That was before the invention of radio, airplanes, television, computers, etc.—and it was 38 years before the invention of xerography. It was 7 years before Chester Floyd Carlson, the inventor of xerography, was born. Carlson's invention marked the beginning of successful commercial applications and systematic fundamental investigations of amorphous semiconductors.

During the last decades, the increasing use of amorphous chalcogenides and VIb group chemical elements of the periodic table—namely selenium, sulfur, and tellurium—in semiconductor films has been very impressive. The following paragraphs explain their relevant properties.

First, the properties of amorphous selenium (a-Se) are well documented and as such, it can be considered as the most representative among this class of materials. In addition, it can serve as an ideal test material for the comparison of various mobility-lifetime product ($\mu\tau$) measurement techniques.

Second, the nature of the deep traps in a-Se and its technologically important chemically modified forms, such as Cl-doped and halogenated a-Se-Te, have not been satisfactorily identified. The capture radius of the deep traps, for example, is not known.

A further reason for using a-Se is that it can be readily prepared by using conventional vacuum-deposition techniques with reproducible properties so that the results presented will be typical for any photoreceptor-grade a-Se or Cl-doped a-Se:As film.

Undoubtedly, a-Se not only offers advantages of reduced cost, but it can also be readily produced as large-area elements of the type required in applications.

The successful use of noncrystalline chalcogenide semiconductors in various applications, especially in various imaging applications, depends upon the development of our understanding of their origin and unique physical properties. These properties can be compared to the standard that presently exists in the case of their crystalline counterparts.

For many years, during and after the development of the modern band theory of electronic conduction in crystalline solids, amorphous materials were not considered for use as semiconductors. The occurrence of bands of allowed electronic energy states, separated by forbidden ranges of energy, had become firmly identified with the interaction of an electronic waveform with a periodic lattice. Thus, it proved difficult for physicists to contemplate the existence of similar features in materials lacking such long-range order.

In hindsight, all of the necessary clues were available, including the ability of conventional glasses to transmit light of sufficiently long wavelengths and the photoconductive behavior of solids like amorphous selenium. The blind spot that had developed was not exposed until the mid-1950s, when Boris Kolomiets and coworkers ushered in the current area of knowledge with reports on semiconducting behavior in various chalcogenide glasses [1].

Stimulated by a variety of commercial applications in the field, such as xerography, solar energy conversion, thin-film active devices, etc., international interest in this subject area increased dramatically after these early reports. The absence of long-range order invalidates the use of simplifying concepts such as the Bloch theorem, for which the counterpart for disordered systems has proved elusive. After more than a decade of concentrated research, there remains no example of an amorphous solid for the energy band structure, and the mode of electronic transport is not a subject for continued controversy.

In contrast to crystalline solids, which are characterized by the long-range periodic order of their constituent atoms, the properties of amorphous solids are determined by their electronic configurations and the chemical bonding of adjacent atoms. These two classes of solids have quite different structures. Freedom from the constraint of atomic periodicity permits a wide range of material compositions to be prepared, which may exhibit insulating, semiconducting, or metallic behaviors. Amorphous insulators have found widespread application in the microelectronics industry, and amorphous metallic alloys have useful magnetic properties. In this book, we restrict ourselves to a discussion of imaging applications and related properties of amorphous chalcogenide semiconductors. There are two groups of amorphous semiconductors that are of greatest commercial interest:

1. The chalcogenide glasses, which contain a considerable proportion of one or more chalcogen elements—selenium, sulfur, and tellurium—often combined with semimetals, such as arsenic or germanium.
2. The tetrahedrally bonded amorphous solids, such as amorphous silicon (a-Si), germanium (a-Ge), and related alloys.

Both these groups may be conveniently prepared in the form of thin/thick films by deposition from the vapor phase onto a suitable substrate. This is of considerable importance in applications where large-area coverage of flat or curved surfaces of rigid or flexible materials is desirable, such as in photovoltaic arrays, X-ray sensors, display screens, and photocopier drums. The energy and material cost involved in producing amorphous films are significantly lower than for comparable crystalline material on the basis of useful area. This is because the slow high-temperature processing involved in producing single-crystal wafers is necessary, and less material is operationally lost. Furthermore, as the structure of amorphous semiconductors is not determined or fixed by thermodynamic equilibrium conditions, it can be changed (sometimes reversibly) by heat, light, or electric field. As a consequence, certain of these materials exhibit unique electronic and optical properties.

Set against these advantages are the limitations caused by the characteristically low carrier mobility that prevails, except under high field and/or high injection conditions. This is likely to preclude the widespread use of amorphous semiconductors for some time to come in electronic devices such as high-speed logic elements. However, amorphous semiconductor–based devices already compete successfully in the marketplace against their crystalline analogs as well as in their own right, and this trend is likely to continue as current developments reach fruition.

I.1 Chronology of Commercial Applications

The first major application of amorphous semiconductors, dating from about 1969, was in xerography ("dry writing"). This process, which was discovered by Carlson in 1938, makes use of the photoconductivity of certain high-resistivity amorphous semiconductors. A thin film of semiconductor is charged positively by a corona discharge. After exposure to light, the surface charge is reduced in approximate proportion to the light intensity by creation of electron-hole pairs. Negatively charged toner particles (carbon black encapsulated in a low melting plastic binder) are attracted to the film and then transferred, using another corona discharge, to a sheet of paper where the image is fixed by heating. Traditionally, single layers of amorphous Se, or As_2Se_3, which is more resistant to light-induced crystallization, have been used in this process. Over the last decades, the xerographic photoreceptors have been progressively using more organic photoconductors rather than selenium alloys. At the same time, some large-volume copying applications still use amorphous selenium alloys since they provide many copies per drum [2]. More recently, multiple-layer structures of hydrogenated amorphous silicon (a-Si:H) have been shown to offer certain advantages: They have good sensitivity in the red and IR regions, long-term image quality, and mechanical durability.

The first commercially available amorphous semiconductor–based electronic device was a nonvolatile digital memory. This was developed following Ovshinsky's discovery of reversible memory switching phenomena in chalcogenide glasses in 1968. Certain chalcogenide glass thin films exhibit electrically controllable, reversible amorphous-crystalline phase transition. The conductivity of the crystalline material is much higher than that of the amorphous material, and it is this property which defines the memory state. It is interesting to note that competition from crystalline silicon-based products, of similar operational specification, has resulted in this device gaining a negligible fraction of the current market share.

Neither of the applications mentioned relies fundamentally upon the use of electronically doped amorphous semiconductors. Until the early 1970s, attempts to achieve doping in these materials had been unsuccessful. It was generally believed that amorphous semiconductors could not be substitutionally doped because impurity atoms would simply be incorporated within the amorphous network at sites that satisfied their valence requirements. Although this view is largely correct, there are, fortunately, some exceptions. In 1969, Chittick and coworkers showed that amorphous silicon films prepared by decomposing silane gas (SiH_4) using an RF field exhibited a large photoconductivity in comparison with films deposited by evaporation. Later, it was shown that these films contained a significant proportion of hydrogen and that improved photoconductivity resulted from reduction in the number of electronic defects. In 1974, workers at RCA Laboratories and Spear and Le Comber at Dundee University showed independently that hydrogenated amorphous silicon could be doped *n*- or *p*-type by adding phosphine (PH_3) or diborane (B_2H_6) to silane prior to deposition. This is also a consequence of the reduction in defect states [3].

These discoveries stimulated a great deal of scientific interest, and the commercial possibilities were recognized. It was not long before reports of various diode structures and rudimentary transistors began to appear in the scientific literature. Photovoltaic elements in particular received a great deal of theoretical and empirical attention, and the drive for improved efficiency and stability continues today. Although a-Si:H photocells are still somewhat less efficient than comparable crystalline semiconductor cells, they can be mass-produced cheaply and thus offer a lower cost per watt of electrical power. These factors have led to them becoming the predominant ambient-light power source for consumer products, such as LCD-based calculators, watches, radios, and cassette recorders. Small-scale power installations, for battery charging and power production in remote location, are becoming increasingly popular.

Personal displays such as televisions screens have been commercially available since 1990. Multiple amorphous layers (a few tens of atoms thick) have found applications as X-ray optical elements, and progress continues to be made at a fundamental level in the understanding of these and similar structures. Many other commercial applications of amorphous silicon have been proposed and are at various stages of development, e.g., image pickup tubes, charge-coupled devices, optical recording media, and electronic memories.

Although exploitation of the properties of chalcogenide glasses has been less spectacular than in the case of amorphous silicon, there are notable exceptions. Photocopiers with amorphous selenium and chalcogenide glass–coated drums are still used, and the Xerox Corporation has developed a color photocopier with a three-layer chalcogenide glass photoreceptor. Hitachi has taken advantage of the photoconductive properties of these materials in an image pickup tube (the Saticon). There are currently a number of potential applications for selenium-based amorphous semiconductors in high-sensitivity TV pickup tubes called the HARP [4], in large-area X-ray sensitive vidicons for medical imaging [5], and in ELIC (electrophotographic light-to-image converter) imaging devices [6]. The structural modification of chalcogenides and chalcogenide-metal films on exposure to light gives rise to a variety of potentially useful phenomena. Crystallization, changes in optical transmission coefficient, and refractive index may be induced, sometimes reversibly, and these effects may find commercial applications in the fields of holography, integrated optics, and optical data storage. Small, well-defined holes can be made in Te-rich chalcogenide glasses using a laser, and there is much current interest in the use of this technique to record high-density digital optical information. Certain metals will dissolve in chalcogenide glasses under illumination and electron-beam irradiation, forming new materials whose optical and chemical properties are changed. As well as providing a means of recording optical images of high contrast and resolution, the resistance of the material to the etching processes used in the semiconductor industry indicates clear possibilities in the field of submicron photolithography.

I.2 Formation and Stability of Amorphous Solids

When an amorphous solid is formed, the constituent atoms must maintain their valence requirements as the disorder of the liquid or precondensed state is "frozen-in" on cooling. Because the average structural relaxation time of the material is a critical function of temperature, cooling must occur sufficiently rapidly and to a low enough temperature for the possibility of crystallization to be avoided. The behavior of a material which may exist in either a crystalline or an amorphous form depends on these conditions.

The ease of formation of an amorphous solid and its subsequent stability are determined by the degree of flexibility inherent in the atomic structure and bonding. In covalent solids, which include the amorphous semiconductors, this may be expressed in terms of average covalent coordination, or connectivity m. This corresponds to the number of bonds a typical atom makes with its neighbors. Elements of group IV, V, and VI, which are the major constituents of amorphous semiconductors, are coordinated four-, three-, and twofold, respectively. The following general rules apply to covalent amorphous materials:

$m < 2$: Structure is insufficiently cross-linked to maintain overall atomic rigidity.
$2 < m < 3$: Structure may accommodate the bonding constrains with only small distortions of bond length angle.
$3 < m < 4$: Structure is overconstrained; broken bond results.

I.3 Atomic Structure

Amorphous chalcogenide semiconductors encompass a wide range of compositions, and their structures and properties vary considerably as a result. As a class of solids, they are often referred to as glasses, a term reserved for amorphous materials whose short-range order is similar in both the liquid and solid states, which may be formed by melt-cooling under appropriate conditions. The average covalent coordination clearly depends on composition, lying typically between two and three.

The elemental chalcogen glass a-Se has a connectivity close to two, and it is believed to possess a predominantly one-dimensional atomic structure consisting of Se_N chains, with $N \approx 10^8$, and possibly some smaller rings such as Se_8. By adding up to 40 at %As, which has a trigonal coordination, a range of binary two-component glasses may be readily obtained by melt-quenching. Glass-forming ability over a wide compositional range is possible because the valence requirements of both the Se and As atoms can be satisfied by relatively strong Se–Se and As–Se bonds without recourse to the weaker As–As bond.

I.4 Electronic Structure

The detailed electronic structure of a given chalcogenide glass is determined by the chemical bonding which occurs and thus by its composition. A primary influence on electronic and optical properties is exerted by the electronic configuration and bonding of the chalcogen atoms. The highest valence band is formed from nonbonding or lone-pair states. In amorphous semiconductors, distortions of bond angle and length occur to a greater extent in comparison with crystalline analogs, and this results in a "tailing" of band edges into a pseudo gap, which contains a finite density of allowed states.

Based on earlier [1–7] and recent [8–14] publications on the topic, the present authors try to give only the most important information about the imaging applications of amorphous chalcogenides, preferentially on Se-based (Se-rich) materials. We follow a logical scheme in presenting the information: preparation (fabrication) of amorphous chalcogenides, molecular structure, electronic defect states in the mobility gap (including spectroscopy of such states), and various imaging applications of these unique materials.

I.5 References

1. B.T. Kolomiets, Phys. Stat. Solidi 7 (1964) 713.
2. S.O. Kasap, in: Handbook of Imaging Materials, eds. A.S. Diamond and D.S. Weiss (Marcel Dekker, Inc., New York, 2nd ed., 2002).
3. D. Adler, Proc. SPIE 617 (1986).
4. K. Tanioka, J. Yamazaki, K. Shidara, K. Taketoshi, T. Hirai, Y. Takasaki, Adv. Electron. Electron. Phys.. 74 (1988) 379.
5. R. Luhta and J.A. Rowlands, in Proc. Conf. Photoelectric Image Devices (London, 1991).
6. K. Kempter, A. Kiendl, W. Muller, H. Voit, J. Non-Cryst. Solids 59–60 (1983) 1219.
7. C.F. Carlson, Electrophotography, US Patent 2, 297, 691 (1942).
8. S.O. Kasap, J.B. Frey, G. Belev, O. Tousignant, H. Mani, L. Laperriere, A. Reznik, J.A. Rowlands, Phys. Status Solidi B 246 (2009) 1794.
9. V.I. Mikla, V.V. Mikla, J. Optoelectron. Adv. Mater. 10 (2008) 131.
10. V.I. Mikla, V.V. Mikla, Optoelectron. Adv. Mater.—Rapid Commun. 1 (2007) 272.
11. M. Popescu, J. Non-Cryst. Solids 352 (2006) 887.
12. G. Lucovsky, M. Popescu, Non-Crystalline Materials for Optoelectronics (INOE Publishers, Bucharest, 2004).
13. A.V. Kolobov, Photo-Induced Metastability in Amorphous Semiconductors (Wiley-VCH, Weinheim, 2003).
14. Y. Ohkawa, K. Miyakawa, T. Matsubara, K. Kikuchi, S. Suzuki, K. Tanioka, M. Kubota, N. Egami, A. Koboyashi, IEICE Electron. Express 6 (2009) 1118.

1 Preparation of Amorphous Selenium Photoconductor Films by Vacuum Deposition

Recently, there has been a revival of interest in pure and selenium (Se)-based amorphous materials due to their commercialized applications in X-ray imaging as X-ray photoconductors. The supersensitive television (TV) pickup tube called the High-gain Avalanche Rushing Amorphous Photoconductor (HARP) is for use in high-definition television (HDTV). The difficulty of producing homogeneous alloy photoreceptor films of selenium has been known since the earliest vacuum deposition works. The electronic quality of amorphous selenium (a-Se) films requires the substrate temperature during vacuum deposition to be as high as possible, above the glass transition temperature. This, in turn, eventually leads to crystallized films, which have relatively high conductivity and are therefore not useful as photoconductors. A variety of stable amorphous chalcogenide materials can be prepared in bulk, thin/thick films, and multilayer forms using melt-quenching, vacuum deposition, and other less common techniques. In this chapter, the main accent is on the conventional method of vacuum deposition, which is not only convenient but also simple and inexpensive. The results show that only the substrate and boat temperatures might be considered as factors determining the required quality of the films. Undoubtedly, the results have a wider significance than a-Se films and can be easily extended to a number of other chalcogenide semiconductor films.

Various unique effects and unusual physical properties inherent to pure and alloyed amorphous chalcogenides—disordered, non-oxide solid state materials containing selenium, sulfur, and tellurium—have not been observed in any other solid state materials, whether they are organic or inorganic.

Amorphous chalcogenides can be prepared by means of several relatively simple and inexpensive methods. Among these, the most exploited are the following:

- Melt-quenching technique used for bulk-glass fabrication.
- Vacuum evaporation with subsequent deposition on different substrates.
- Spin coating.
- Fiber technology.

In addition to these techniques, there are special technological methods, such as radio frequency or laser beam sputtering.

Faced with the question of imaging applications of amorphous chalcogenides, researchers have convincing arguments that thin and/or thick film of amorphous chalcogenides prepared by vacuum evaporation with subsequent deposition on

Amorphous Chalcogenides. DOI: 10.1016/B978-0-12-388429-9.00001-7

substrates are not only potential candidates for digital image formation in medical diagnostics. Today, they are also successfully used in HDTV pickup tubes and media for optical (holographic) storage of a high density of information.

The initial structure, especially in the case of sulfur-based films, depends on the preparation method, although less significantly as for crystalline semiconductors. In fact, the structure of As-prepared amorphous films differs from that of well-annealed films and bulk-glass counterparts of identical chemical composition. This difference is minor in the case of pure and selenium-rich amorphous films.

Because of the importance of recently commercialized applications in X-ray imaging with X-ray photoconductors (see [1] and the review by Kasap and Rowlands [2]), there has been a revival of interest in the growth characteristics of especially thick a-Se films. Amorphous selenium satisfies three conditions necessary for use as an X-ray photoconductor in imaging systems. First, it has a good X-ray photosensitivity, thus producing a high signal. Second, it has very high resistance and this limits dark current. Third, it can be easily and relatively inexpensively prepared by vacuum evaporation in the form of large area thick films with reproducible properties. The large area is necessary for human chest radiography. Amorphous selenium that is used in actual X-ray photoconductor applications is not simply pure but stabilized. The stability is obtained by alloying a-Se with a small amount of arsenic (As) and doping it with chlorine (Cl) in the parts per million range. The role of these two additional components is extremely important: (a) the addition of As (which cross-links Se chains) greatly retards crystallization of pure a-Se (this process takes place even at room temperature); (b) doping with Cl is necessary to prevent the action of deep hole traps produced by the addition of As. The recent Toravision digital X-ray imaging system, which was developed by Philips Corporation, uses a thick a-Se photoconductor film coated on an Al drum [3]. The basic principle of operation of such a photoconductor is similar to xeroradiography [4]. In xeroradiography, the system photocopies a body part by using X-rays. At the same time, there is no powder development. The electrostatic image readout system is based on rotation of the photoconductor drum under a line of electrostatic voltage probes to read the electrostatic voltage.

Currently, a-Se is also used in high-sensitivity pickup tubes called vidicon-type High-gain Avalanche Rushing Amorphous Photoconductor device HARPICONs for HDTV [5] in which an avalanche of photogenerated holes leads to an optical gain and a high sensitivity which is similar to that in the avalanche photodiode.

It is a well-known fact that a-Se alloys, including a-As_2Se_3, have been successfully used as xerographic photoreceptors for several decades. Indeed, this was a multibillion dollar copy industry technique. However, over the last decades, these conventional materials have been replaced by much cheaper organic photoconductors [6].

The physical properties of amorphous semiconductors are much less sensitive to the preparation conditions than those of their crystalline counterparts. Nevertheless, there are a number of parameters that can potentially influence molecular structure and electronic properties of pure a-Se films prepared by thermal evaporation in a vacuum. These are the following: the boat temperature, the evaporation rate, and the substrate temperature. In addition, various post-deposition treatments, such as annealing and illumination with band-gap light, among others, can modify the sample properties.

Electronic quality a-Se films, with good electron and hole lifetimes, require the substrate temperature during vacuum deposition to be as high as possible, above the glass transition temperature. But this eventually leads to crystallized or polycrystalline films, which have high conductivity and are therefore not useful as photoconductors. In this paragraph, we consider the key relationship between the substrate temperature T_{sb}, the boat temperature T_b, and the properties of the resultant a-Se films. The vacuum deposition of a-Se film onto a substrate will produce either an amorphous or polycrystalline film. Keck [7] discovered that the substrate temperatures below 50°C produced an amorphous selenium film while a substrate in excess of 90°C would cause many small hexagonal selenium crystallites to appear in the film.

Givens [8] found that the amorphous form of selenium was relatively easy to create by deposition and could be subsequently converted to a crystalline form by heating the film above 85°C. This suggests that there is a given substrate temperature beyond which the amorphous film phase cannot be obtained. It is well known that high substrate temperatures result in polycrystalline Se films. At the same time, it should be noted that the majority of articles on vacuum coated a-Se films published earlier (during the 1970s and 1980s) reveal remarkable little on the details of the vacuum-coating conditions. Therefore, the empirical relationship between the boat and substrate temperatures remains unclear. The only exceptions are probably the results reported by Montrimas et al. [9]. They have observed that as the substrate temperature is increased above 50°C, the amorphous films contain trigonal (hexagonal) crystalline inclusions whose density increases with the substrate temperature. There exists another example. The thin film formation and growth occurs through absorption of impinging selenium atoms on stable clusters on the substrate and the subsequent growth of these clusters as the evaporation continues. It is not clear, however, how these thin film growth characteristics influence the structure and properties of thick films. Thick films are prepared by high coating rates from open boats and by increasing the substrate temperature results in small crystalline inclusions in an amorphous matrix.

Juhasz et al. [1] explain the experimental relationship—$T_{bt} = f(T_{sb})$—for the borderline between a-Se films with and without significant crystalline inclusions using the following model: crystalline inclusions occur at heterogeneous nucleation sites on the substrate. Their growth to significance within the amorphous phase is a rate competition process between the rate of crystal growth, assumed to be inversely proportional to the viscosity, and the material condensation rate which controls the growth of the amorphous phase. The crystal growth rate depends on the substrate temperature, and the material condensation rate depends on the boat temperature.

1.1 Preparation of Amorphous Se Films for Imaging Applications

Amorphous Se film samples are usually prepared by thermal evaporation of selenium pellets from a molybdenum boat in a standard vacuum system. Typical vacuum pressure is less than 10^{-5} torr. Both the source and substrate temperatures are

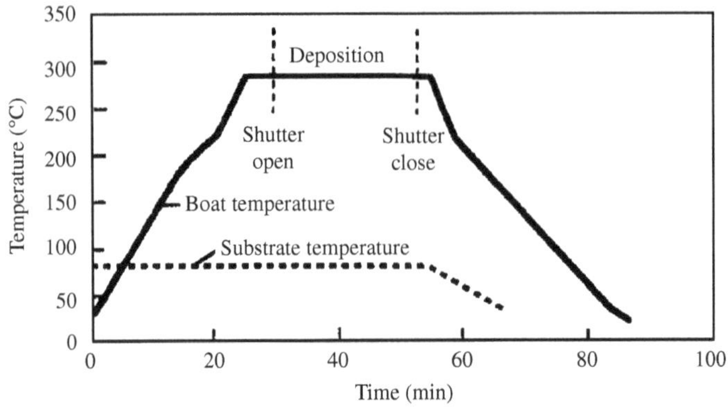

Figure 1.1 Typical boat and substrate temperature versus time profile. The boat temperature and substrate temperature are controlled and constantly maintained throughout the deposition [1].

electronically controlled. Figure 1.1 shows a typical boat and substrate temperature versus time profile.

A shutter function protects the substrates until steady evaporation conditions have been established and also during cooldown following deposition. For depositions using boat temperature lower than the melting point of selenium, the pellets are first melted at 230°C and then cooled to the desired temperature. Various samples are prepared with the boat temperature T_b kept constant at a value in the range of 190–250°C. The substrate temperature is typically 60–65°C, above the glass transition temperature of pure a-Se. Most samples are prepared with substrate temperature T_s in the range shown previously and lead to electronic quality a-Se films.

On the other hand, the boat temperature T_b influences the deposition in two ways. First, T_b simply controls the rate of deposition. Second, T_b determines the composition. In other words, T_b determines the distributions of different molecular species in the vapor phase. For selenium, increasing T_b from 190°C to 250°C results in Se_5 increasing, and this species dominate the gas phase [6]. The pellets used for the starting material are either high-purity selenium or high-purity selenium alloyed with various concentrations of arsenic and chlorine in the case of so-called stabilized amorphous selenium. The concentration of particular vacuum deposited films of stabilized amorphous selenium will be somewhat different than the bulk (source) material due to the effect of fractionation.

A radiative heater (as the substrate heater) was used and its temperature could be maintained within a 0.5°C specification [1]. For example, the 821 source controller allowed sample temperature–time profiles to be programmed based on separate ramp or dwell sections within one evaporation cycle. Thermocouples are attached to the source and substrate. As one can see from Figure 1.1, both the boat and the substrate temperatures were maintained constant during the deposition process by appropriately shuttering the evaporation. For each sample, one or two initial trial and error runs established the shuttering points to maintain T_b constant during deposition and result in a 60-μm film. At $T_b = 280°C$, a 60-μm film was deposited in 24 min, resulting in an average deposition rate of about $2.5\,\mu m\,min^{-1}$.

A stainless steel boat with two carriage heaters was used to evaporate the source material. A load of about 15 g of vitreous Se pellets would give a film of 60 μm on a substrate at 15 cm from the source. These Se pellets were standard industrial production batches. A shutter was used to prevent deposition on the substrate during the temperature ramp of the boat and to allow deposition when the boat reached the required temperature. All source temperature readings for the controller were taken from a thermocouple suspended in the source cavity in the boat amongst the pellets just above the floor.

Identical aluminum sheets were used as substrates. The supplied sheets were mechanically polished and sealed in plastic. In order to ensure that the surface was flat and bore no traces of either the polishing compound or the plastic film, each substrate was thoroughly cleaned. The reader may find more details on this procedure in Juhasz et al. [1].

Next, we consider the experimental observations and results without introducing a specific model for a-Se film growth. A number of selenium films of approximately 60 μm thickness were deposited for various pairs of values for T_{sb} and T_b. All other conditions, such as the source–substrate distance, background pressure, boat load, shuttering points, and substrate size, were kept the same. The polycrystalline films were easily recognizable, as they have a bright, textured structure and typically occur at high substrate temperatures. Amorphous films appear as a uniform black glass layer—a-Se below its glass transition temperature (50–60°C) is essentially a brittle glass. A number of amorphous looking films prepared at elevated substrate temperatures showed visible spots that were identified as crystalline regions or crystalline inclusions. These crystalline regions tended to appear at heterogeneous nucleation sites on the substrate. Eventually, within about a 2–3°C further rise in T_{sb}, the films became polycrystalline. Reasonably, the amorphous films with crystalline inclusions are not usable as device-quality a-Se photoconductor films. These general observations are in agreement with the qualitative descriptions of similar experimental work carried out by Montrimas et al. [9].

In his work, Keck describe the method for examining crystallinity [7]. The deposited films were immersed in a CS_2 solution, which is known to dissolve the amorphous structure but leave the hexagonal crystalline form intact. Amorphous films prepared at low (60°C) substrate temperature were totally dissolved in CS_2, which indicated that these films had no crystalline regions. All these films were labeled as A (Figure 1.2), representing the amorphous phase. Amorphous films prepared at high substrate temperature, when dissolved in CS_2, left behind undissolved small crystals on the substrates, typically about 1 mm in lateral dimension. The number of undissolved crystals left on the substrate was counted, and an average crystal density N_{cr} was calculated as the number of crystals per unit centimeter square of the substrate area. The films were then categorized as amorphous and labeled as A if $N_{cr} = 0$, amorphous with crystalline inclusions and labeled as AC if $N_{cr} = 2–4\,cm^{-2}$, and amorphous with excessive crystalline inclusions labeled as C if $N_{cr} > 4\,cm^{-2}$. A 1–2°C rise in T_{sb} for AC films was sufficient to lead to a polycrystalline sample.

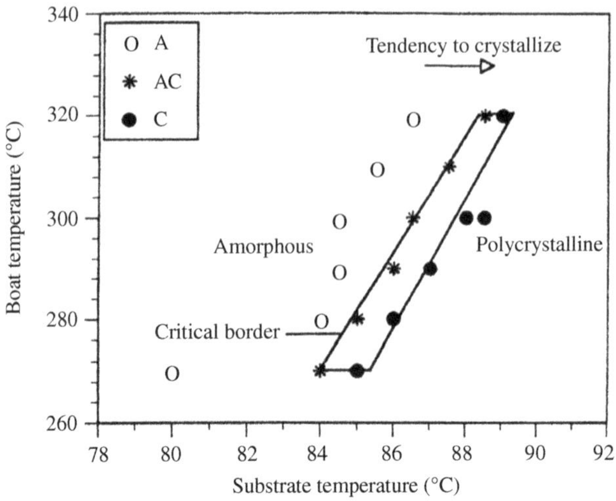

Figure 1.2 Boat temperature versus substrate temperature for samples A (amorphous), AC (amorphous with only small number of crystalline inclusions), and C (samples with significant crystalline inclusions) [1].

Figure 1.2 shows the results of the experiments as T_b versus T_{sb} for a given range of N_{cr}. As Juhasz et al. stated, an immediate feature is the appearance of a well-defined boundary region between amorphous films (A films) and those films with crystalline inclusions (AC films and C films) [1]. Although the T_b–T_{sb} boundary is relatively steep, it is not vertical; that is, T_{sb} is not independent of T_b. There seems to be a clear relationship between T_b and T_{sb}, which is discussed later. Higher boat temperatures tend to result in boat splitting, and lower temperatures result in deposition rates that are too low to be practical. Figure 1.2 constitutes a selenium film structure diagram in terms of practical deposition variables T_b–T_{sb} for the vacuum coating of a-Se films. It is quite likely that the border region between amorphous and crystalline films will also depend on other vacuum system parameters, such as the source–substrate distance—this affects the deposition rate and is equivalent to changing the source temperature.

It is necessary to accentuate that there are two reasons for analyzing the experimental data in terms of the boat temperature rather than the film deposition rate. First, a-Se-based photoreceptor films are prepared by monitoring the boat temperature (rather than monitoring the exact deposition rate during the coating process). The thickness monitor tends to identify the beginning and end of the evaporation process and any unusual occurrences during coating. Second, the rate of deposition cannot be monitored accurately because the pinhole in the quartz crystal case slowly becomes clogged. Not using a pinhole results in an excessive amount of material condensation during the evaporation, which causes the crystal to go out of oscillation.

1.2 Model for Amorphous–Crystalline Film Boundary

Crystallization of a-Se films was studied extensively [8]. Typically, the crystallization of a-Se films occurs by heterogeneous nucleation. The latter process means that the overall crystallization rate is primarily determined by the rate of crystal growth. Since the crystal growth rate is generally inversely proportional to the viscosity, η, the temperature dependence of the crystallization rate follows that of the reciprocal viscosity. It is necessary to note here that selenium-based chalcogenide glasses do not generally exhibit a viscosity with an Arrhenius temperature dependence. Crystallization experiments carried out at different temperatures indicate that the activation energy E_{cr} measured for the crystallization process depends on the temperature. A more appropriate representation is to describe the crystallization rate by

$$R_{\text{crystallization}} \propto \frac{1}{\eta} \propto \exp\left(-\frac{A}{T - T_0}\right), \tag{1.1}$$

where it is assumed that the viscosity is characterized by a Vogel–Tamman–Fulcher (VTF) behavior: A and T_0 are the VTF parameters. This type of VTF crystallization rate has been shown to describe the crystallization data reasonably well over a wide temperature range encompassing the glass transformation as well. Although the VTF behavior can be readily related to free-volume theories, it is difficult to relate it simply to an Eyring transformation with a well-defined energy barrier.

The crystal growth rate initiated at various heterogeneous sites in the deposited film on the substrate T_{sb} is then given by

$$R_{\text{crystallization}} \equiv B\exp\left(-\frac{A}{T_{sb} - T_0}\right), \tag{1.2}$$

where B is a constant. It is assumed that the film temperature is uniform and equal to the substrate temperature. The crystallization proceeds by heterogeneous nucleation from the substrate. It is also clear in Figure 1.2 that the temperature range of interest is T_{sb}, which is only 5–6 K; hence, we are considering crystal growth over a narrow temperature range. One can therefore represent the VTF behavior by a simple Arrhenius rate without loss of accuracy as

$$R_{\text{crystallization}} \equiv C'\exp\left(-\frac{E_{cr}}{kT_{sb}}\right), \tag{1.3}$$

where C' is a constant, k is the Boltzmann constant, and E_{cr} is the apparent activation energy that is related to A by

$$\frac{1}{E_{cr}^{1/2}} = \frac{1}{(kA)^{1/2}} - \frac{1}{(kA)^{1/2}}\left(\frac{T_0}{T}\right). \tag{1.4}$$

As assumed in [1], E_{cr} can be determined by using an average temperature for T in Eq. (1.4).

The rate of evaporation of source material from the boat depends on the boat temperature and can be represented by an Arrhenius rate of the form

$$R_{\text{evaporation}} = K' \exp\left(-\frac{\Delta H_{\text{ev}}}{RT_{\text{b}}}\right), \tag{1.5}$$

where K' is a constant, R is the gas constant, ΔH_{ev} is the evaporation enthalpy per mole, and K has an algebraic temperature dependence ($T^{-1/2}$) that can be neglected compared with the exponential term. The selenium condensation rate will be proportional to the evaporation rate. The condensation rate on the substrate decreases with distance L of the substrate from the boat. This distance is constant for a given vacuum coater geometry. Thus, we can write the condensation rate as

$$R_{\text{condensation}} = K \exp\left(-\frac{E_{\text{ev}}}{kT_{\text{b}}}\right), \tag{1.6}$$

where K is a constant (which is proportional to $1/L^2$) and we have converted ΔH_{ev} to E_{ev}, the evaporation enthalpy per atom. The rate of film growth will be the same as the condensation rate on the substrate.

The condensed material will be amorphous and crystallizing at selected favorable heterogeneous substrate sites at a rate of $R_{\text{crystallization}}$. In the case of $R_{\text{crystallization}} < R_{\text{condensation}}$, the crystallization rate is too slow compared with the condensation rate to allow a heterogeneously nucleated crystallite to grow into a significant crystalline inclusion. Thus, the ratio $R_{\text{condensation}}/R_{\text{crystallization}}$ determines crystallinity/amorphicity of the deposited film. The critical boundary between crystalline–amorphous films is when the two competing rates are comparable

$$C \exp\left(-\frac{E_{\text{cr}}}{kT_{\text{sb}}} \approx \exp\left(-\frac{E_{\text{ev}}}{kT_{\text{b}}}\right)\right), \tag{1.7}$$

where $C = C'/K$. After rearranging, we have a T_{sb}–T_{b} relationship that defines the critical boundary between crystalline and amorphous films

$$T_{\text{b}} \approx \frac{E_{\text{ev}}/k}{\ln[C \exp(E_{\text{cr}}/kT_{\text{b}})]}. \tag{1.8}$$

The constant C can be determined by choosing one point on the experimental critical boundary. Equation (1.8) would then predict the $T_{\text{sb}} = f(T_{\text{b}})$ boundary passing through the above-mentioned point. This boundary can then be compared with the experiments to evaluate the model. Thus, the model is examined by evaluating the closeness of Eq. (1.8) to the whole T_{b}–T_{sb} experimental data.

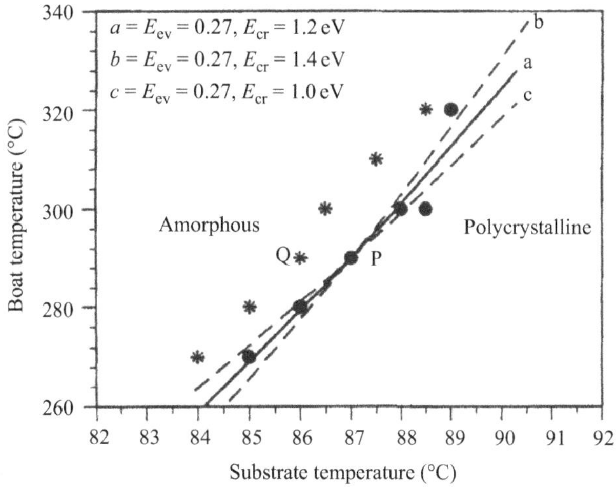

Figure 1.3 Theoretical boat temperature versus substrate temperature curves (*a*, *b*, and *c*) compared with the experiments [1]. The corresponding symbols are defined in Figure 1.2.

Figure 1.3 shows as solid and dashed lines, labeled *a*, *b*, and *c*, the calculations based on Eq. (1.8) using $E_{ev} = 0.27\,eV\,atom^{-1}$ and three different values for E_{cr} to encompass all possible values in this temperature range. The predicted curve *a* uses $E_{cr} = 1.2\,eV\,atom^{-1}$, which is based on A and T_0 values determined for nearly all the crystallization studies performed on a-Se. Curve *b* uses $E_{cr} = 1.4\,eV$. Curve *c* uses a lower value of $1\,eV\,atom^{-1}$ as an average for E_{cr}, as reported for a-Se films crystallizing at higher temperatures [9–11]. It is remarkable that Eq. (1.8) with accepted values for E_{ev} and E_{cr} can almost totally account for the curvature of the experimental T_b–T_{sb} behavior. The theory predicts a T_b–T_{sb} behavior represented by curve *a* lying between the curves *b* and *c*. One can see that this region contains the experimental points. The model proposed by Juhasz et al. [1] provides a reasonable description of the observed T_b–T_{sb} behavior and can be extended, with appropriate modifications, to describe the crystallization behavior of various a-Se alloys and other chalcogenide semiconductors.

It is important to note that the agreement between the theoretical curve in Figure 1.3 and the experimental data is quite sensitive to the relative choices of E_{cr} and E_{ev}. The curvature of the T_b–T_{sb} curve is sensitive to the ratio of E_{cr} to E_{ev}, which is about 4.44. The spread *b*–*c* is a 20% variation in either E_{cr} or E_{ev}, given this relative ratio.

Finally, it should be emphasized that an important assumption of the model proposed in [1] is that as soon as the substrate temperature is sufficiently high, those favorable heterogeneous nucleation sites on the substrate with the lowest nucleation activation energy enable the nucleation and growth of crystalline regions which then appear as spots on the final a-Se film. As the substrate temperature is increased, more and more nucleation sites are activated, N_{cr} increases, and eventually, at high enough substrate temperatures, there are sufficiently close crystallites to result in the whole film becoming polycrystalline.

It is instructive to give here a short summary of the above results with a variety of thick a-Se films prepared on Al substrates by conventional vacuum deposition when only the substrate and boat temperature were varied. Experiments indicated the appearance of crystallites in the amorphous Se film as the substrate temperature T_{sb} was increased. A critical T_b–T_{sb} boundary between amorphous films with and without crystalline inclusions was identified. This model has been developed for the critical relationship between T_{sb} and T_b, that is, $T_b = f(T_{sb})$ based on two competing rate mechanisms: the rate of deposition (or material condensation on the substrate) and rate of crystal growth in the film assuming heterogeneous nucleation at favorable substrate sites. Over the narrow substrate temperature range, Juhasz et al. [1] approximated the VTF behavior of the crystal growth rate by an Arrhenius rate with apparent activation energy E_{cr} that accounts for most crystallization studies on a-Se. The condensation rate was assumed to follow an Arrhenius rate with the activation energy determined by the enthalpy of evaporation, E_{ev}. The $T_b = f(T_{sb})$ model was particularly sensitive to the E_{cr}/E_{ev} ratio. By using the presently accepted values for E_{cr} and E_{ev}, it was shown that the experimental results could be fitted to the $T_b = f(T_{sb})$ model. This model has a wider significance than a-Se and can be easily extended for other chalcogenide semiconductor films.

1.3 Influence of Deposition Conditions on Electronic Properties of Amorphous Selenium

Figure 1.4 summarizes the mobility and lifetime values as functions of T_b. As seen in this figure, the lifetime and mobility of both electrons and holes are relatively independent of T_b, irrespective of the deposition rate variation by two orders of magnitude. The lack of any change in the electronic properties at conditions when the vapor phase changes indicates that all arriving species are either decomposed or incorporated into the growing film in a uniform way regardless of the particular vapor phase composition. So if T_s is maintained above the glass transition temperature during the deposition process, the surface atomic mobility of deposited species enhances, and this allows them to find the optimal positions.

On the other hand, the deposition of a-Se at room temperature leads to film samples with poor electronic properties. Subsequent annealing of these samples partially improves their electronic properties, but they cannot compete with the properties of films obtained at temperature near glass transition temperature. At first glance, the experimental fact that electronic properties of selenium are not affected by the Se vapor composition seems to be a signature of independence of a-Se structure on the vapor phase molecular species [10]. The authors assume (based on their own experimental results) that, although the molecular species which are present in vapor phase depend on T_b, the structure of deposited films after annealing procedure becomes identical to that of bulk glass.

The structure of a-Se is discussed further in Chapter 2, and here we note only that it is believed to consist of chainlike and some ringlike regions. Chainlike Se regions are fragments of Se_n chains and have good hole transport but poor electron transport.

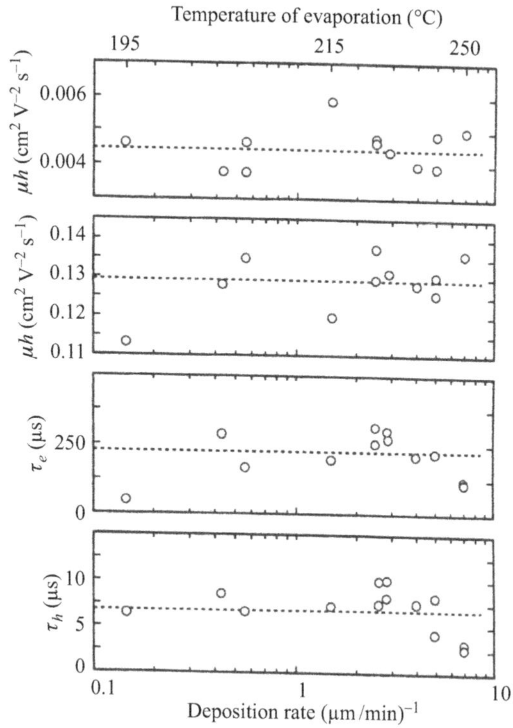

Figure 1.4 The influence of evaporation temperature and deposition rate on electronic properties of a-Se [10].

Ringlike regions, which are fragments of Se_8 rings, have good electron transport but exhibit poor hole transport.

Electronic properties of a-Se do not depend on the selenium vapor composition. Therefore, our previous assumption that the structure of deposited a-Se films is not sensitive on the vapor phase molecular species seemed to be valid.

The effect of As and Cl addition on the transport properties of a-Se has been studied extensively by means of xerographic and time-of-flight (TOF) techniques during the last decade (see [12] for references). Xerographic experiments were well described by Abkowitz [13]. They involve corona charging of the sample in darkness to some initial voltage V_0, and after that, the following measurements: dark discharge rate dV/dt, first cycle residual potential V_{r1} (the surface potential which remains after photodischarge of corona-charged sample), and the saturated residual potential V_{rn} in repeating charge–photodischarge cycles. The residual potential V_{r1} in the first cycle gives the $\mu\tau$ product values. Xerographic measurements provide $\mu\tau$ products only for holes because a-Se cannot be charged negatively to high voltages. The deep trapping time can be determined from conventional TOF experiments by decreasing the electric field until the carrier transit time $t_T = L/(\mu E)$ becomes comparable with the lifetime τ. There are several disadvantages to the technique when it is applied to a-Se-based alloys. These are the following:

- Photogeneration depends on the electric field. Thus, reduction of the field too much results in significant decreasing of the photocurrent signal and, reasonably, long lifetimes cannot be measured.

- The composition of the Se-based alloys is not uniform across the sample. So the photocurrent shape will be strongly affected by the fractionation effect.
- The photocurrent waveform is influenced by the presence of bulk space charge. As a result, a nonuniform field profile has much more of an effect on photocurrent shape than trapping.

All these effects make the determination of deep trapping time from conventional TOF experiments unambiguous.

In contrast, the interrupted field time-of-flight (IFTOF) technique [14] can provide realistic values for τ. The IFTOF measurement involves interrupting the transit of photoinjected charge carriers during their flight across the specimen in the conventional TOF measurement. The interruption is achieved by removing the applied field at time t_1. After an interruption period of t_i, the field is reapplied at time $t_2 = t_1 + t_i$, and the photocurrent is redetected. Figure 1.5A shows the conventional TOF experiment. The resulting nearly trap-free TOF photocurrent is shown in Figure 1.5B; Figure 1.5C displays the interruption of the bias voltage at time t_1 until time t_2, when it is reapplied. The ideal interrupted TOF photocurrent in which displacement currents are absent is displayed in Figure 1.5D as a result of the application of the bias in Figure 1.5C. Figure 1.5C and D also define the various time marks used in the remainder of the extensive article [14]. The lifetime τ is obtained from the slope

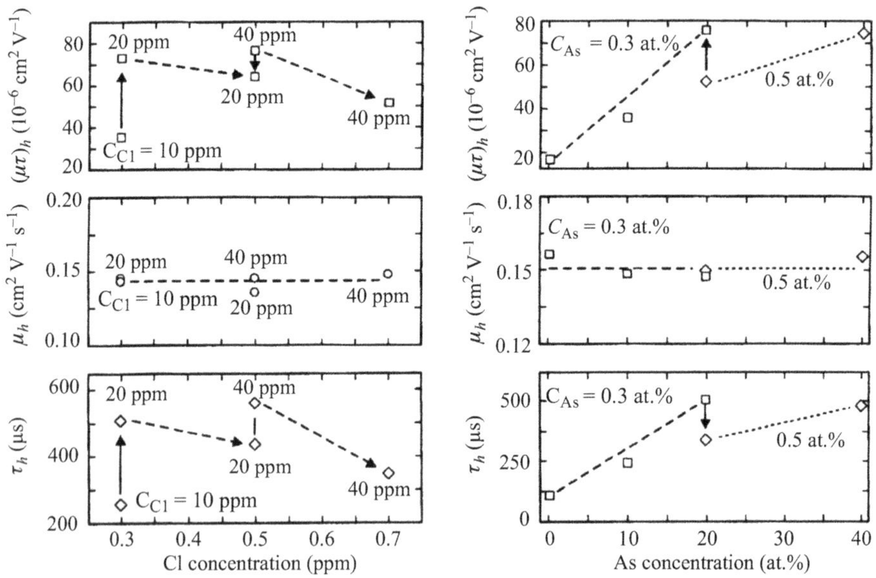

Figure 1.5 The influence of As and Cl on hole lifetime, mobility and hole range in a-Se. Schematic illustration of IFTOF technique principle. (A) The conventional TOF measurement technique. V_A is the applied voltage, and X_A and X_N are triggers for the applied voltage and the xenon flash. (B) Conventional TOF photocurrent signal. Δ is due to the dispersion broadening of the photoinjected charge sheet; T_t is the transit time. (C) The IFTOF applied voltage profile, $V_A(t)$. (D) IFTOF photocurrent signal resulting from the application of the bias, $V_A(t)$ in (C). T_t is the mean time of arrival of the photoinjected holes, which also includes the interruption time t_i [14,15].

of the line (i_2/i_1) versus (t_2-t_1) plot. The IFTOF technique has been widely used by Kasap who obtained accurate τ values for both electrons and holes in a wide range of Se-based alloys [16].

The same article critically reviews the first time the IFTOF principle, with its various distinct advantages, was applied to the measurement of the spatial dependence of the hole lifetimes in stabilized a-Se (As- and Cl-doped amorphous Se) X-ray plates used in X-ray imaging. The hole lifetime could be measured as a function of location in the term, and the changes in the spatial variation of the lifetime could be determined upon exposure to X-rays. The IFTOF technique is shown to be an extremely powerful tool for studying spatial dependence of charge transport and trapping parameters in the sample.

It is clear from Figure 1.6 that the hole lifetime and the hole range both decrease with As addition. At the same time, as the Cl content is increased, there is a

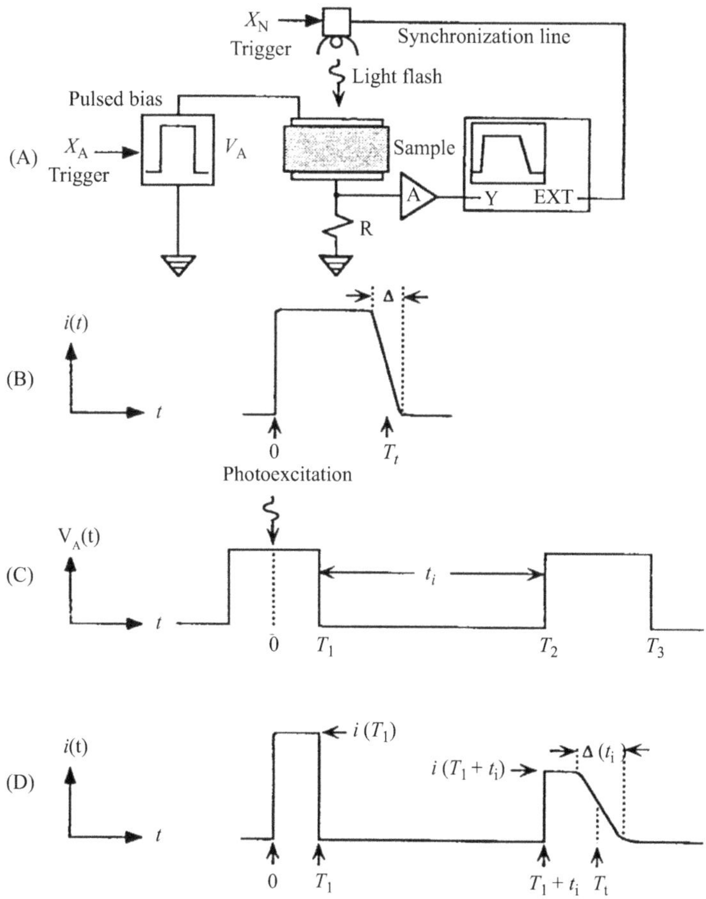

Figure 1.6 The influence of As addition and Cl doping on hole lifetime, mobility, and hole range in a-Se [10].

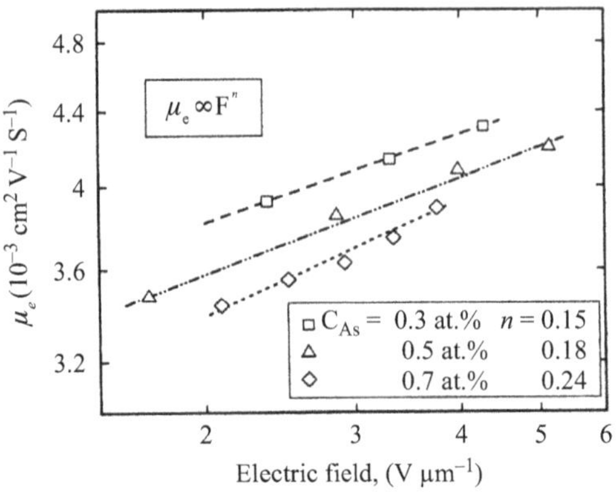

Figure 1.7. The electron mobility as a function of electric field F applied to a-Se–As–Cl [10].

corresponding increase in τ and $\mu\tau$. The As and Cl addition has an opposite effect on hole lifetime and range. The hole drift mobility seems to be unaffected by either As or Cl addition. Therefore, we can control the hole range by appropriately choosing the relative amounts of As and Cl.

It is interesting, on the other hand, to examine the influence of As and Cl on the electron lifetime and mobility. The electron mobility μ_e in a-Se depends on the applied electric field [17]. In order to get comparable values for different samples, the dependence of μ_e versus electric field was measured (Figure 1.7) and determined for the values corresponding to the same electric field (4 V μm^{-1}).

Figure 1.8 shows the effects of As and Cl on electron transport parameters. It is clear that the electron lifetime τ_e behaves in opposite manner as the hole lifetime τ_h. Increasing the concentration of As increases the electron lifetime, whereas increasing Cl content decreases the electron lifetime. As the arsenic content is increased, the electron mobility tends to decrease. It is necessary to note that the increase in the lifetime is greater, therefore the electron range $\mu\tau$ increases with As addition, and this seems to be a distinct technological advantage. At the same time, Cl does not affect electron mobility (Figure 1.8).

One can compare qualitatively the effect of As and Cl addition on transport parameters. For similar large changes to be observed in the corresponding properties, only small amounts of Cl and larger amounts of As are needed (ppm range for Cl and 0.1–1.0% for As). As an example, the reduction in electron lifetime caused by increasing As content up to 0.5 at.% may be compensated by the addition of only 20 ppm of Cl.

1.4 Fractionation Effects in Amorphous Se–Te Films

Many of the properties of amorphous $Se_{1-x}Te_x$ semiconductor alloys have been studied during the 1970s and 1980s [6,12,18–20]. Recently, there has been a revival of

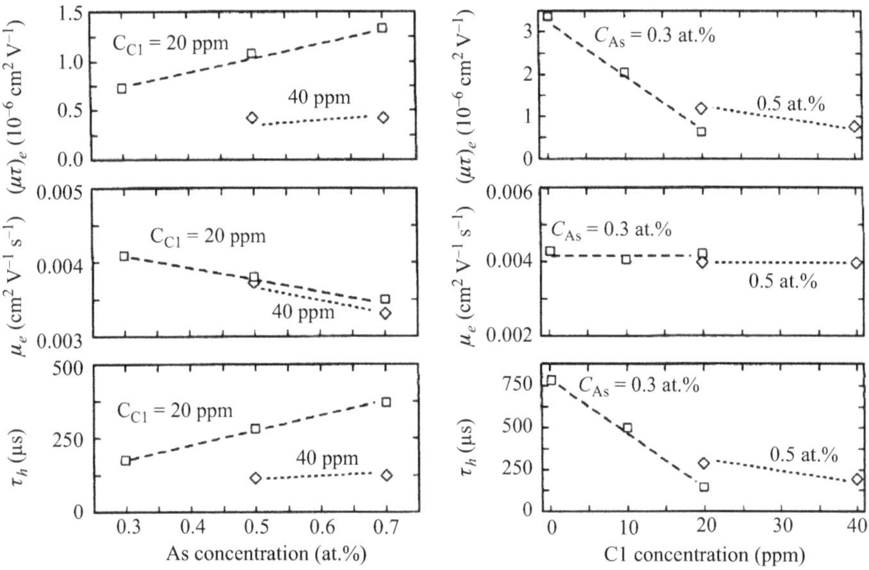

Figure 1.8. The influence of As and Cl addition on electron lifetime, mobility, and electron range in a-Se.

interest in Se–Te photoconductors since Tanioka and coworkers have been successful in fabricating a commercial supersensitive TV pickup tube called the HARPICON for use in HDTV [21,22]. There is difficulty in producing homogeneous alloy photoreceptor films of amorphous selenium [12,23]. The reader may find the results of systematic study and explanation of fractionation effects in the case of vacuum-coated a-Se–Te photoconductor films in [24]. There is an illusion in the selenium-depleted nature of the film surface with a higher than expected concentration of Te (see [24] for references). The variation of composition across the photoconductor film thickness is referred to as *fractionation*, and it leads to a variation in the physical properties across the photoconductor film. The study of IFTOF clearly shows this tendency [14,15]. It should be noted that no published work has sought to explain the fractionation phenomena or to provide a detailed investigation of its operation under various deposition regimes. The only exception is [24].

A typical plot of Te concentration across the thickness of an a-Se–Te photoreceptor film (thickness of about 60 μm) is shown in Figure 1.9. The presence of a discernable Te-rich region, a plateau of relatively uniform bulk, and a Se-rich zone are clearly seen in this figure. Juhasz and coworkers performed an experimental study of fractionation effects in vacuum-coated a-Se–Te films and proposed a model of the evaporation process to account for these effects [24]. The reader can find in this extensive work a systematic study of the fractionation in a-Se–Te films during conventional vacuum deposition (using an open stainless steel boat) for various batches of Te:12.5 wt% source material. The fractionation model is based on present

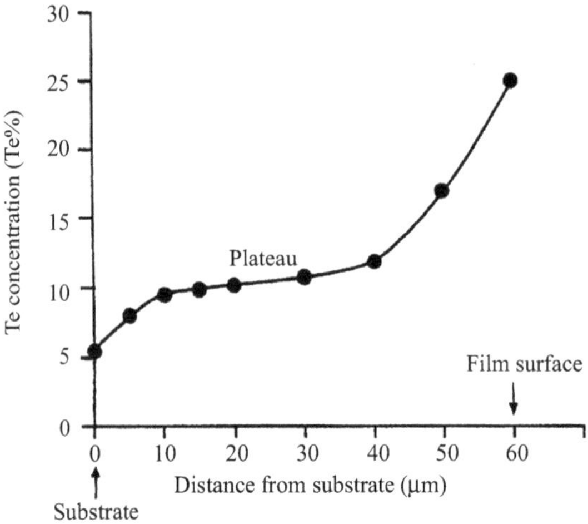

Figure 1.9 Concentration of Te across a typical a-Se–Te photoconductor film of thickness 60 μm. The source composition was Te:12.5 wt%. The boat temperature at the evaporation was 280°C [24].

understanding of the molecular species in the Se–Te vapor and how Te is likely to enter the molecular species in the vapor from the surface of the molten alloy in the boat.

During the initial stage of the evaporation process, more Te is left behind in the melt because not all Te atoms can enter the molecular species in the vapor. The Se–Te alloy in the boat vaporizes incongruently. The vapor is initially Se-rich, and this fact explains the existence of the Se-rich zone near the substrate. The diffusion of Se and Te atoms to and from the surface region (SR) of the melt is unable to homogenize the whole source material. Therefore, the SR of the melt has a higher Te concentration than the bulk. The Te content of the melt surface increases until a dynamic quasi-equilibrium is reached. During this process, the diffusion and flux of Se and Te atoms from the bulk to the surface is just balanced due to evaporation from the melt surface. This fact can explain the existence of the plateau region in Figure 1.9. As the evaporation process proceeds, the melt becomes too rich in Te, and the diffusion and flux from the bulk at this stage is unable to maintain the quasi-equilibrium. The vapor becomes Te-rich, which leads to a Te-rich SR in the deposited film. Although this model is only semiquantitative, it explains the general features of the experimental results.

Charge transport properties, namely, hole and electron drift mobility, in vacuum-coated a-Se–Te films have been reported by various authors in the past using the conventional TOF transient photoconductivity technique. The hole mobility μ_h decreases rapidly with the Te content (up to ~5% Te) when a-Se is alloyed with Te. The decrease in the mobility becomes small beyond this concentration of Te. A minimum in μ_h versus Te content dependence was observed by Vaezi-Nejad and Juhasz and Abkowitz [26–31] with coworker at about 8–10% Te [25,26]. The hole mobility has been found to depend on the electric field E, as $\mu_h \propto E^n$, where n is a small index of order $(1\text{–}2) \times 10^{-1}$ that depends on the Te content. Although the fractionation has undesirable effects in the case of vacuum-coated photoreceptor films, it is difficult to avoid this effect. Only a small portion of the Te concentration is suitable for homogenous film preparation (Figure 1.9). This is in contrast to the distance from the substrate, which corresponds to the flat plateau region. Hole and electron transport at room temperature were re-examined in homogeneous films. The drift mobility and the hole lifetime were measured by using the IFTOF transient photoconductivity technique.

It is useful to describe the essential details of the preparation of a-Se-based alloy photoreceptor films here. Film samples used in commercially available devices were prepared in a stainless steel conventional vacuum chamber pumped by an oil diffusion pump with a liquid nitrogen cold trap (Figure 1.10).

The typical vacuum pressure was approximately 10^{-3} Pa. Both the source and the substrate temperature were controlled during evaporation. The substrate heater was a radiative heater. To prevent deposition on the substrate during the temperature ramp of the boat and to allow deposition when the boat reached the required temperature, as in the cases of most Se-based amorphous materials, a shutter was used. In such a situation, one can eliminate the bombardment of the substrate by imperfectly vaporized segments of evaporant. Another positive role of the shutter (when used appropriately) is to prevent the impurities, which may have collected at the bottom of the source, to access the substrate.

The source material was in the form of small, vitreous pellets. So only 12.5% composition was examined for the fractional studies—this is a typical photoreceptor composition.

Juhasz et al. [24] have proposed a theoretical model for Se–Te fractionation. The fact that Te concentration is not homogeneous in the vacuum-coated films suggests a time-dependent relationship between the Se and the Te atoms that are present in the melt. To understand such a phenomenon, certain initial assumptions were made. The most essential element is the existence of close correlation between the Te concentration of the sample surface and that in the vapor (in other words, the composition of the molten alloy surface at the source).

It can be proposed that a Te-rich SR builds up due to an imbalance between the processes of flux and diffusion acting on Se and Te atoms in the bulk. This refers to the whole melt excluding the thin SR. This situation is shown in Figure 1.11 as a vertical slice through the source boat at the condition when the contents are fully molten and the evaporation is in progress. This Te-rich SR may be considered to be a cornerstone of the model, and it has been verified experimentally by examining a slice of source material by suddenly terminating the evaporation at some point in the deposition process.

The theory was further developed by applying the basic mass transport concept to the SR. The SR is assumed to contain a constant number of atoms and has a nearly constant thickness Δz. The thickness of the SR is much smaller than the whole thickness of the boat material. Let us suppose that N_{ev} is the rate of evaporation of atoms per unit area from the *SR*. N_f and N_d are the number of atoms moving per unit time per unit surface area from the bulk region into the surface by flux and diffusion, respectively. Then during evaporation, in the SR

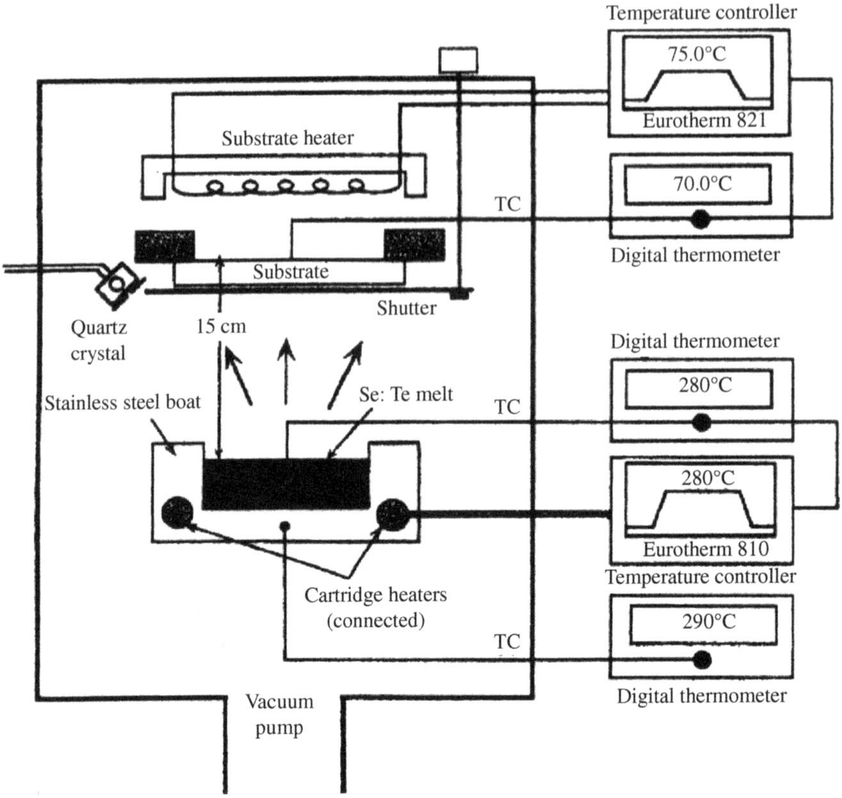

Figure 1.10 The vacuum coating system usually used for preparation of a-Se–Te films [24].

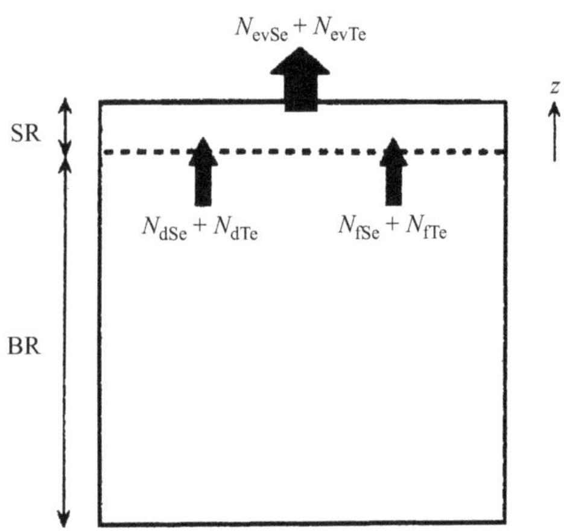

Figure 1.11 Schematic illustration of the fractionation model, which represents a cross-section of the source material in the boat [24].

$$N_{ev} = N_f + N_d.$$

The modeling of the evaporation process proposed in [24] involves obtaining the composition of the SR as the evaporation proceeds from beginning to end. We can determine the composition of the vapor and hence the deposited film at that instant by considering the allowed molecular species in the vapor.

It is apparent from Figure 1.12 that one can prepare homogeneous a-Se–Te films. This purpose may be realized by allowing the substrate to capture only the plateau region by properly shuttering the substrate during evaporation.

Carrier transport in a-Se–Te films was investigated by using conventional TOF and IFTOF transient photoconductivity measurements. There are two essential features that should be mentioned. First, hole transport is measurable up to the highest Te concentration of 15%. Second, the electron transport signal is present at low Te concentration (~4%). The drift mobility is determined using a well-known formula $\mu = L^2/(t_T V)$. In this equation, L is the film thickness and V is the applied voltage.

Hole and electron drift mobility for a-Se is 3% Te, and 3.5% Te depends on external electric field application Figure 1.13. This dependence is shown in Figure 1.13 and follows $\mu \sim E^n$ type behavior, in which n increases with Te content.

The authors agreed with the interpretation of Juhasz et al. [24] that this type of behavior represents a true charge transport effect.

The dependence of hole drift mobility and lifetime in a range of a-Se–Te films doped with Cl in the 10–20 ppm amount is shown in Figure 1.14. The Cl doping reduces the density of deep traps in the mobility gap [25]. The variation of μ_h

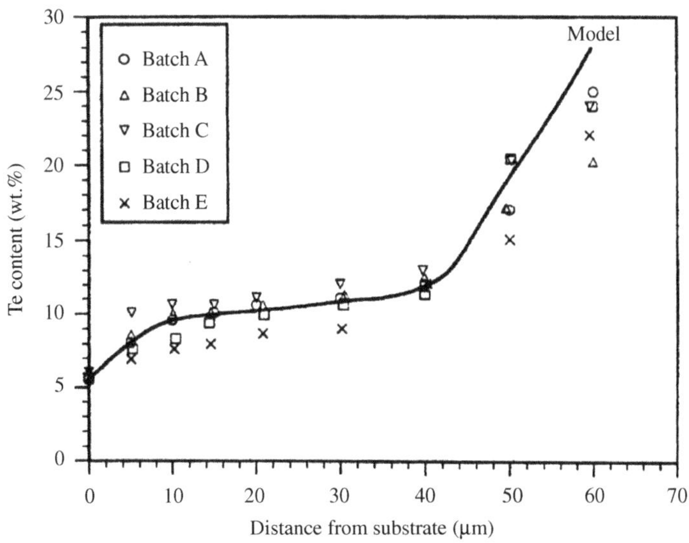

Figure 1.12 Results of the fractionation experiments and the model (only selected batches, which include extremes of deviation from the model, are shown) [24].

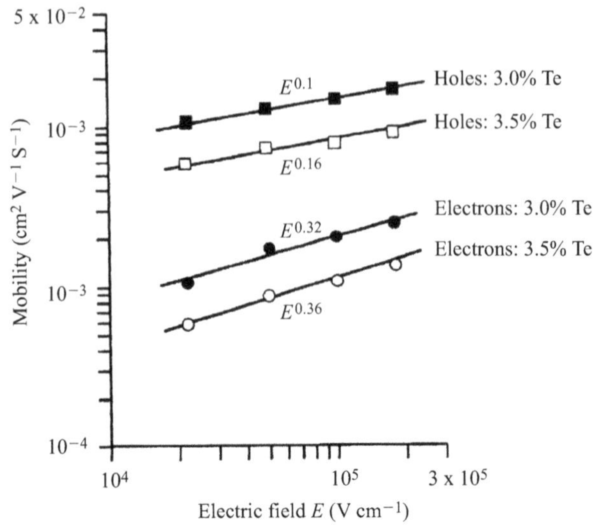

Figure 1.13 Hole and electron drift mobility versus electric field in a-Se–Te homogeneous films.

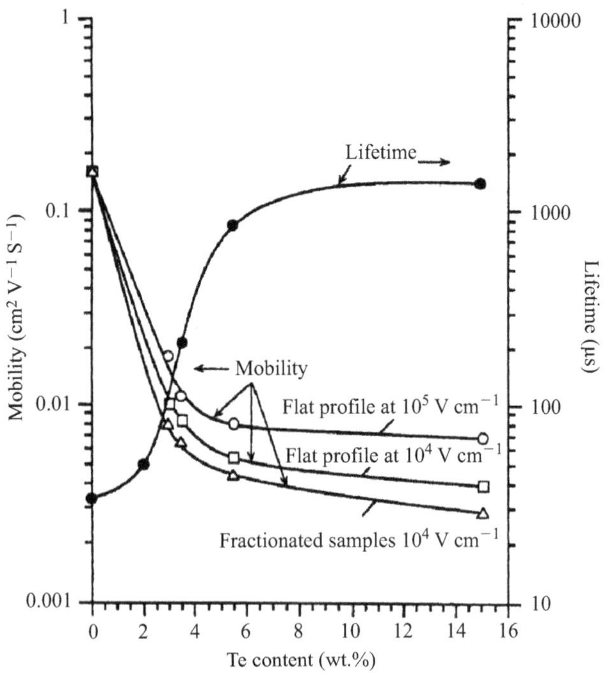

Figure 1.14. Hole drift mobility and hole lifetime in Cl-doped a-Se–Te films. Also shown are the results for homogeneous and fractionated films.

with the Te content follows the previously reported behavior, and no minimum in this dependence is observed, as reported in [26]. In contrast, the lifetime τ_h increases with Te content and exhibits an opposite behavior compared with the drift mobility μ_h. As a result, the $\mu_h\tau_h$ product (in other words, the hole range) remains relatively constant and insensitive to variations of Te content. This behavior may be interpreted to mean that Te introduces a set of shallow traps that are not situated at the same energy as intrinsic shallow traps in a-Se [27]. The drift mobility reduces as the concentration of shallow traps increases, and this causes an increase in the lifetime. At the same time, with Cl doping, the concentration of deep hole traps remains practically unchanged.

1.5 Conclusion

Experimental results obtained show that variation of substrate and boat temperatures during conventional vacuum deposition are the only factors determining the electronic quality of pure and alloyed amorphous Se-based films. These films are used for imaging applications as X-ray photoconductors and high-sensitive TV pickup tubes.

1.6 References

1. C. Juhasz, V. Gembala and S.O. Kasap, J. Vac. Sci. Technol. A 18 (2000) 665.
2. S.O. Kasap and J.A. Rowlands, J. Mater. Sci. Mater. Electron. 11 (2000) 179.
3. U. Neitzel et al., Med. Phys. 21 (1994) 509.
4. J.W. Boag, Phys. Med. Biol. 18 (1973) 1.
5. K. Tsui et al., Mater. Res. Soc. Symp. Proc. 219 (1991) 505.
6. B.E. Springet, in Proc. of the Fifth International Symposium on Uses of Selenium and Tellurium (Selenium-Tellurium Development Ass., Inc.) Grinbergen, Belgium (1994) 187.
7. P.H. Keck, J. Opt. Soc. Am. 42 (1952) 221.
8. M.P. Givens, Rev. Sci. Instrum. 25 (1954) 1130.
9. E. Montrimas, B. Petretis, J. Bakauskas, and B. Baliukas, Lietuvos. Fisikos Rinkinyas 13 (1973) 273.
10. S.O. Kasap et al., Russ. Phys. Techn. Semicond. 37 (2003) 816.
11. V.I. Mikla, Doctor of Sciences Thesis, Academy of Sciences, Kiev(1998).
12. S.O. Kasap, in Handbook of Imaging Materials (eds., A.S. Diamond and D.S. Weiss) 2nd ed. (New York, Basel, Marcel Dekker) (2002) 329.
13. M. Abkowitz and R.C. Enck, Phys. Rev. B 25 (1982) 2567.
14. S.O. Kasap, R.P. Thakur, and D. Dodds, J. Phys. E. Sci. Instrum. 21 (1988) 1195.
15. S.O. Kasap, B. Polischuk, and D. Dodds, Rev. Sci. Instrum. 61 (1990) 2080.
16. S.O. Kasap, B. Polischuk, Can. J. Phys. 7 (1995) 96.
17. C. Juhasz, S.M. Vaezi-Nejad, and S.O. Kasap, J. Imaging Sci. 29 (1985) 144.
18. I. Cheung, G.M.T. Foley, P. Fournia and B.E. Springett, Photogr. Sci. Eng. 26 (1982) 245.
19. B.E. Springet, in Proc. of the Third International Symposium on Industrial Uses of Selenium and Tellurium (Selenium-Tellurium Development Ass., Inc.) Grinbergen, Belgium (1984) 258.

20. B.E. Springet, in Proc. of the Fourth International Symposium on Uses of Selenium and Tellurium (Selenium-Tellurium Development Ass., Inc.) Grinbergen, Belgium (1984) 126.
21. K. Tanioka et al., Adv. Electron. Phys. 74 (1988) 379.
22. K. Tsuji et al., Mater. Res. Soc. Symp. Proc. 219 (1991) 505.
23. J.S. Berkes, in Electrophotography: Second International Conference (ed., D.R. White) Society of Photographic Sciences and Engineering, Springfield, VA, 1974, p. 137.
24. C. Juhasz, V. Gembala, and S.O. Kasap, J. Mater. Sci.: Mater. Electron. 10 (1999) 633.
25. Y. Wang, C.H. Champness, J. Appl. Phys. 77 (1995) 722.
26. S.M. Vezi-Nejad and C. Juhasz, Thin Solid Films 148 (1987) 251.
27. S.O. Kasap and C. Juhasz, J. Non-Cryst. Solids 72 (1985) 23.
28. J.C. Schottmiller, J. Vac. Sci. Technol. 12 (1975) 807.
29. S.M. Vaezi-Nejad and C. Juhasz, Semicond. Sci. Technol. 2 (1987) 809.
30. S.O. Kasap and C. Juhasz, J. Mater. Sci. 21 (1986) 1329.
31. M. Abkowitz and I.M. Markovics, Solid State Commun. 44 (1982) 1431.

2 Molecular Structure of Se-Rich Amorphous Films

Molecular structure and its transformation are examined for amorphous Se-rich As_xSe_{1-x} ($0 \leq x \leq 0.2$) alloys by employment of diffraction and non-diffraction structural probes. The molecular structure of amorphous Se on the scale of short-range order is close to that of the crystalline phase, while medium-range order differs from the structure of most inorganic glasses and may be placed between three-dimensional network glasses and polymeric ones. Further experiments show the existence of successive phases in laser-induced glass-crystalline transition with pronounced threshold behavior. Below the energy density threshold, E_{th}, only small changes in the local structure of the system can be detected. Above E_{th}, the changes are attributed to crystallization transformation. The corresponding Raman spectra reveal transformation of the system from amorphous into the crystalline phase under laser irradiation. In the binary As_xSe_{1-x} glass system, a change of structural regime takes place near the composition $x \approx 0.04$. The presence of this topological threshold is established by direct and indirect evidence, such as peculiarities in the composition dependence of the basic parameters for electron diffraction and Raman vibrational modes. The peculiarities are caused by the transition from a chain-ringlike structure to a chainlike structure.

Amorphous selenium has been extensively studied over the past 70 years, but its molecular structure is still disputable. For a long time, it was believed that the amorphous phase consisted of a selenium chain Se_n and 8-ring Se_8 structures mixed together. This model arose from the fact that in the crystalline phase, selenium can exist in two forms, α-monoclinic Se (α-Se) and trigonal Se (γ-Se). The former consisted of Se_8 rings and the latter Se_n chains. Reasonably, one can consider a structure for the amorphous phase based on a mixture of ring and chain members. However, the length of the selenium chains is uncertain. In addition, it is unclear whether a-Se contains cyclic structure or rings.

In recent years, there has been enhanced interest in structural studies of chalcogenide glasses owing to their imaging applications. As-containing chalcogenides, especially Se-rich alloys, have usually been studied in thin film form. Change in the molecular structure of charge carriers, which transport, trap, and carry out other important fundamental properties, can have an influence on photogeneration, so knowledge of these materials is needed for further improvement of their characteristics [1–11].

Amorphous Chalcogenides. DOI: 10.1016/B978-0-12-388429-9.00002-9

Undoubtedly, elemental semiconductors are useful tools for studying the influence of structure on physical properties [5–7,12–19]. This is particularly true of selenium. In the past, selenium was successfully used in photocells, rectifier diodes, and solar cells. In its amorphous form, selenium has a good application as a photoreceptor in copying machines and X-ray imaging plates [1,8,11].

Commercial use of a-Se in various applications is increasing (see [8] for references), such as in X-ray flat panel detectors for medical purposes [1]. But its structure is still not fully understood.

Binary noncrystalline As–Se system semiconductors are also of continued scientific and practical interest because of the potential in their technological uses (e.g., as functional elements of multilayer photoreceptors in xerography). Among them, the stoichiometric composition As_2Se_3 and compositions from the range 30–50 at.% As are perhaps the most studied [12–19]. Most stable binary glasses in As–Se system atomic ratios can be varied in a wide range. Although the information about various physical properties of Se-rich alloys is not so extensive and numerous [5–8,15–20], their compositional dependence manifests extrema, or thresholds, in the range of 6–12 at.% As. It is necessary to accentuate that the As–Se amorphous alloy system displays the main extrema of various properties at the stoichiometric composition, where the valence requirements appear to be satisfied. It is reasonable to connect the mentioned non-monotonic behavior with the specific character of local structure changes.

The studied materials consist of certain type of glasses that have chalcogen atoms, sulfur, selenium, and tellurium and exhibit various photoinduced phenomena (the reader may refer to Shimakawa [10] and Tanaka [21]). Among these are photostructural transformations and photocrystallization phenomena, where a change in optical, electrical, and other physical properties is observed. The phase transformation of selenium and its alloys can also be induced relatively simply by laser illumination [22–26]. Reasonably, this unique property makes selenium attractive for use in optical data storage and holographic recording. Many experimental results using selenium and its alloys have been reported [27], but few cases of phase transformation properties have been mentioned.

In this chapter, electron diffraction and Raman scattering in pure amorphous selenium and Se-rich As–Se amorphous films are studied. We will clarify the structural transformations induced by light treatment and compositional changes, focusing mainly on photocrystallization transformations. This chapter also deals with the composition-induced structural modifications in amorphous As_xSe_{1-x}. As seen in the analysis of electron diffraction and Raman scattering data, some discontinuity of atomic arrangement exists with rising As content.

2.1 Techniques Exploited in Structural Studies

Methods used for structural study of solids may be classified into two groups:

1. Direct methods include diffraction of X-rays, electrons, and neutrons.
2. Among indirect methods, Raman scattering, infrared absorption, extended X-ray absorption fine spectroscopy, and nuclear magnetic resonance can be considered.

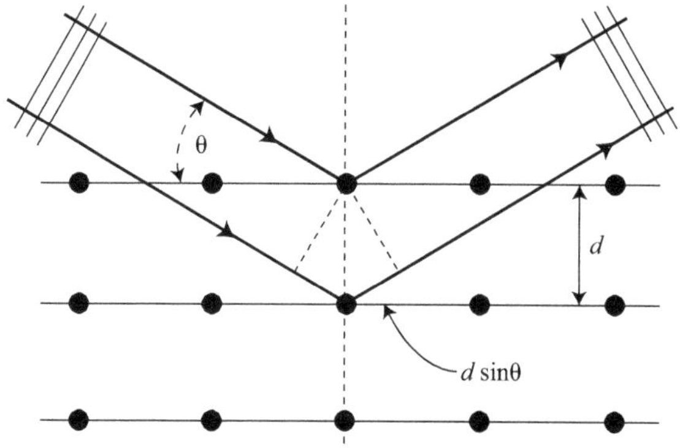

Figure 2.1 Illustration of X-ray diffraction from solids. The incoming beam causes each atom to reradiate a small portion of its intensity as a spherical wave. If atoms are arranged symmetrically with a separation d, these spherical waves will only be in directions where their path-length difference $2d\sin\theta$ equals an integer multiple of the wavelength λ. In that case, part of the incoming beam is deflected by an angle 2θ, producing a reflection spot in the diffraction pattern.

The direct methods determine the arrangement of atoms within a solid. A beam of X-rays strikes a crystal and diffracts into many specific directions (Figure 2.1). From the angles and intensities of these diffracted beams, a crystallographer can extract information about the mean positions of the atoms in the solid, as well as their chemical bonds and disorder.

Since many materials are solids—such as semiconductors, as well as various inorganic, organic, and biological molecules—X-ray crystallography has been fundamental in the development of many scientific fields. In its first decades of use, this method determined the size of atoms, the lengths and types of chemical bonds, and the atomic-scale differences among various materials, especially alloys. X-ray crystallography is still the chief method for characterizing the atomic structure of new materials.

It should be noted here that while diffraction of X-rays, electrons, or neutrons gives definite information about the structure of crystalline solids, such measurements give much less information about the structure of amorphous solids [28,29]. The diffraction patterns of amorphous solids consist of diffuse rings whose radial variation of intensity provide merely a one-dimensional representation of the three-dimensional glass structure. Some information about the structure is averaged out in the experiments, and the structure is surely not one-to-one with the diffraction data. In practice, one imagines a likely amorphous structure, makes a prediction, and compares with the observation.

This illustrates the need for the use of many different experimental probes. Diffraction, vibrational spectroscopy, and other techniques emphasize different aspects of structure, such as interatomic distance, angles, local symmetry, or range of order. Each technique provides an alternative view of the structure and generally

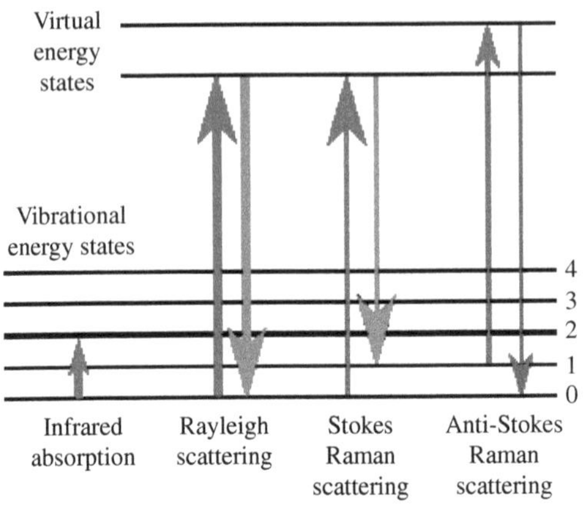

Figure 2.2 Visible light scattering from molecules.

averages the structure differently, thus revealing aspects that may be de-emphasized or obscured by another technique. Although each probe gives limited information, the results of several different probes lead to a convincing conclusion.

Raman spectroscopy (named after C. V. Raman) is a spectroscopic technique used to study vibrational, rotational, and other low-frequency modes in a system. It relies on inelastic scattering of monochromatic light, usually from a laser in the visible, near infrared, or near ultraviolet range. The laser light interacts with phonons or other excitations in the system, resulting in the energy of the laser photons being shifted up or down (Figure 2.2). The shift in energy gives information about the phonon modes in the system.

Raman scattering is known to be a powerful technique, which provides information about the local structure of amorphous chalcogenides [30,31].

2.2 Effect of Composition on Structure of As_xSe_{1-x} Amorphous Films—Electron Diffraction Study

In the following experimental results of electron diffraction and electron microscopic measurements of the structure of As_xSe_{1-x}, amorphous films are considered [27,32]. The results of structural investigations of films with low As content have been compared with the literature data.

The samples used in these studies were amorphous films, about 10 μm thick, prepared by vacuum thermal evaporation of the powdered As_xSe_{1-x} melt-quenched material at the rate of 1 μm min^{-1}. Freshly cleaved KCl single crystals on the (001) plane were used as substrates. The As_xSe_{1-x} bulk glasses were prepared according to the conventional melt-quenching method by agitating the ampoule content for homogenization. Annealing of the films was carried out in air at ambient pressure and at temperatures 40°C, 45°C, 50°C, 60°C, and 70°C for 0, 2, 5, 10, and 20 at.% As, respectively. The composition of the films was determined by electron probe microanalysis, and the compositions quoted are accurate to within ±0.5 at.%. Thin film samples were kept in complete darkness until measured to minimize exposure to light sources, which could lead to changes in the properties and structure of the films. The thermal evaporation method was used due to the necessity of reproducing films with the chemical composition of glasses. Moreover, the evaporation of As–Se glasses is not of a dissociate character, but it proceeds with fractionation [4]. The temperature T_{ev} of an open-type tantalum evaporator, measured with Pt/Pt–Rh thermocouple, was within $670 \leq T \leq 770$ K, depending on glass composition.

The chemical composition of amorphous films was checked by the X-ray spectral microprobe method using the EMMA-4 instrument with a relative error of ±1%.

The atomic structure of a film and its changes with thermal annealing were investigated by the electron diffraction method using the EMP-100 instrument, and the microstructure was investigated in the EMB-100 B transmission electron microscope. It is necessary to note here that the film thickness for structural investigations did not exceed ~600–700 Å. The short-range order parameters of the atomic film structure were obtained by calculating the radial distribution function (RDF). In the dependence $I = f(s)$, where $s = 4\pi \sin\theta/\lambda_e$, θ, is the diffraction angle and λ_e is the wavelength of electrons, and this was plotted in a direct electron current recording mode.

The data about the character and changes in the medium-range order of the atomic film structure were measured by the analysis of positions and intensity of the first sharp diffraction peak (FSDP). The thermal annealing of films was carried out in situ in the electron diffractometer column.

X-ray spectral microprobe data of the film composition was the same as that of the As_xSe_{1-x} initial glasses (within the measurement error). This similarity showed that quasi-equilibrium evaporation conditions were maintained.

The electron diffraction patterns for As_xSe_{100-x} As-deposited amorphous films at $x = 0$, 2, 4, 6, 8, 10, and 40 are given in Figure 2.3. For comparative analysis, we plotted I/I_2 intensity curves, where I_2 is the intensity of the second peak of the $I(s)$ curve, which is less affected by the natural aging of films [32].

As seen in Figure 2.3, the presence of FSDP in the region $s_{FSDP} \approx 1 - 1.5$ is characteristic of diffraction patterns for As-deposited films, and it is associated with the medium-range order of the structure [33–35].

The data on the analysis of the FSDP position, s_{FSDP}, and the second peak, s_2, and their intensities, I_{FSDP} and I_2, for different compositions of As_xSe_{100-x} amorphous films is given in Figure 2.4. Here, rounded and square symbols represent the distribution in values of data (for measurements on films prepared with the same evaporation parameters) for compositions with $x = 0$, 5, and 10. As seen in Figure 2.4, the reproducibility of the parameters, s_{FSDP}, s_2, I_{FSDP}, and I_2 for different evaporation conditions for films of the same composition is comparatively good.

After thermal annealing of films in situ (in the electron diffraction column), we observed a structural transformation, $A_1 \rightarrow A_2$, with retention of the amorphous state, where A_1 and A_2 were the phase state before and after transformation, respectively. Upon further heating, the films crystallized.

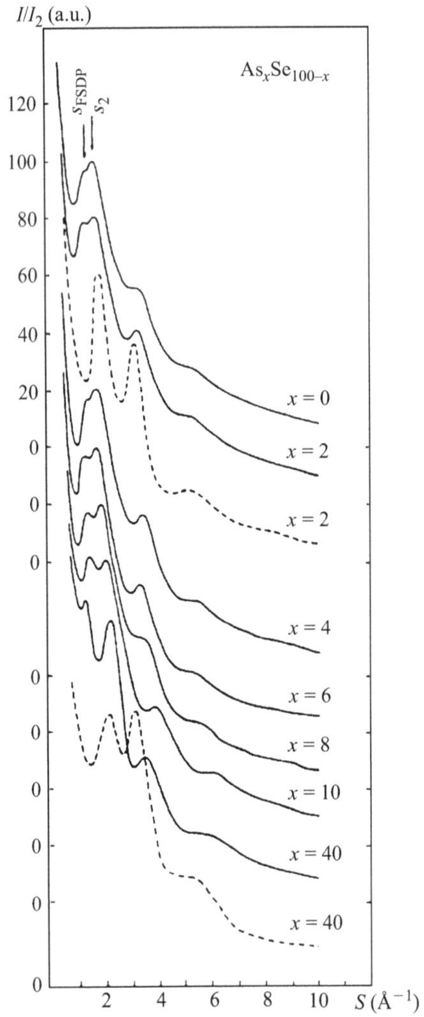

Figure 2.3 Electron diffraction patterns of As_xSe_{100-x} As-deposited amorphous thin films (solid line) and films after structural transformation (dashed line), stimulated by thermal annealing.

The $A_1 \rightarrow A_2$ structural transformation in amorphous chalcogenide films is accompanied by a drastic change in the absorption contrast of the electron microscopic image. The main difference of diffraction patterns between films in the A_2 and in the A_1 state lies in a complete disappearance of T_0 transition for the FSDP, I_{FSDP}, at a certain temperature. In Figure 2.3, typical diffraction patterns of $As_{40}Se_{60}$ and As_2Se_{98} amorphous films after the $A_1 \rightarrow A_2$ structural transformation are shown by dashed lines. New values of s_2 and I_2 are shown in Figure 2.4 as triangular symbols. These values come from diffraction patterns of As_xSe_{100-x} amorphous films of different compositions after $A_1 \rightarrow A_2$ transformation.

It should be noted that diffraction patterns of As_xSe_{100-x} films with low As content after annealing are very similar to those of Se bulk glassy samples. This agrees with

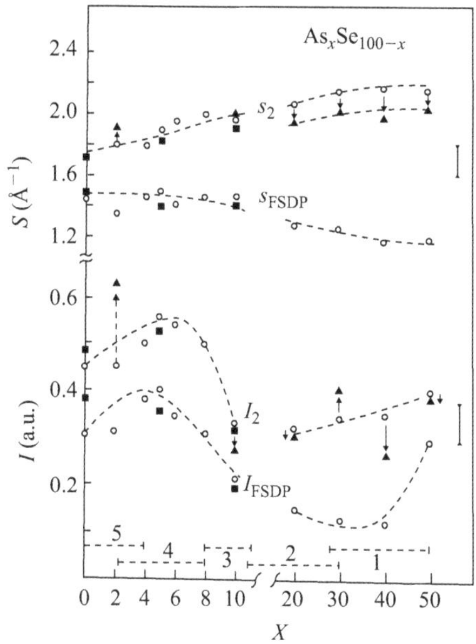

Figure 2.4 Compositional dependence of s_{FSDP}, s_2, I_{FSDP}, and I_2 for amorphous As_xSe_{100-x} films. Solid circles and squares refer to different deposition conditions; solid triangles and open symbols refer to s_2 and I_2 after $A_1 \rightarrow A_2$ structural transformation. Lines are visual guides.

the results of [31,36], in which authors showed that the structure of As_xSe_{100-x} deposited films after annealing approached that of corresponding glasses.

After thermal annealing of As_xSe_{100-x} amorphous films with compositions $x \leq 10$, the $A_1 \rightarrow A_2$ structural transformation occurs even with slight heating. The heating comes from the electron beam of instrument, and films are then crystallized. The recording of diffraction patterns for such films in the amorphous state after $A_1 \rightarrow A_2$ transformation is difficult.

In Figure 2.5, the RDFs of As_xSe_{100-x} amorphous films for $x \leq 10$ are given, and in Figure 2.6, the data on the values of the first RDF maximum position, r_1, and area under it, z_1, are brought together. The values of r_1 and z_1 for films with $x = 30$, 40, and 50 are given for comparison. The designation of the experimental points in Figure 2.6 corresponds to similar designations in Figure 2.4.

The medium-range order of the structure of amorphous films may be characterized to a certain extent by values of s_{FSDP} and I_{FSDP}. From the analysis of Figure 2.4, it follows that As_xSe_{100-x} As-deposited films in the region of compositions $0 \leq x \leq 10$ and $20 \leq x \leq 50$ differ in correlation length, $L = 2\pi/s_{FSDP}$, of the medium-range order and in the degree of medium-range ordering.

The FSDPs for amorphous films of composition close to $As_{40}Se_{60}$ are explained in the approximation of the cluster model. According to this model, the correlation length L is associated with the size of quasi-molecular (cluster) species in the structure of As-deposited films [37].

In the region of low As content, the calculated values of correlation length are $0.42 \leq L \leq 0.46\,\text{nm}$, whereas for the region of compositions $30 \leq x \leq 50$, one obtains $0.50 \leq L \leq 0.55\,\text{nm}$. The latter is fairly well fitted [38] with the distance

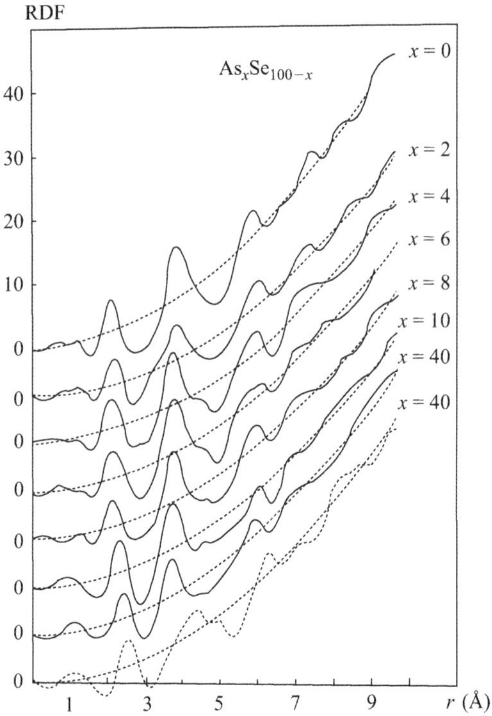

Figure 2.5 Radial distribution functions of As_xSe_{100-x} As-deposited amorphous films (solid lines) and films after structural transformation (dashed lines), stimulated by thermal annealing.

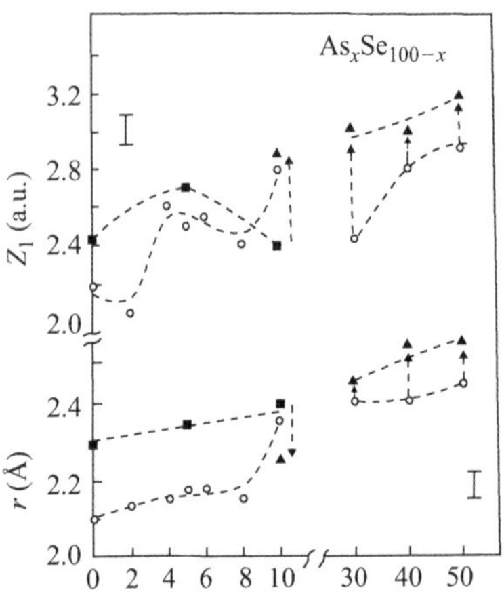

Figure 2.6 Compositional dependence of the first RDF maximum position, r_1, and area under it, z_1, for As_xSe_{100-x} amorphous films.

between the furthest Se atoms in As_nSe_m quasi-molecular species ($n = 4$, $m = 6$). The average value of L in the compositional range $0 \leq x \leq 10$ correlates with the sizes of Se_6 molecules and with the average distance between Se_n chains in the structure of amorphous selenium [4].

We suppose that the peculiarities observed in the compositional dependence of structural characteristics for As_xSe_{100-x} films with low As content are conditioned by a topological structure of the disordered network. These peculiarities appear in those local regions where linking and branching of chains and formation of closed rings occur. In practice, this results in an increase in the intensity of the main diffraction peak I_2, where peak I_{FSDP} is responsible for the medium-range order and a mutual change in the position of these peaks occurs.

2.3 Raman Scattering in Pure and Alloyed Amorphous Selenium: High-Frequency Spectral Region

Raman scattering is a very powerful experimental technique for providing information on the constituent structural units in a given material [30]. In the present section, the Raman scattering in pure amorphous selenium (a-Se) and Se-rich As–Se amorphous films is studied. Our discussion clarifies the structural transformations induced by light treatment and compositional changes. We focus our attention mainly on photocrystallization transformations. In addition, the composition-induced structural modifications in amorphous As_xSe_{1-x} are analyzed. As seen in the detailed analysis of Raman data, with rising As content, some discontinuity of atomic arrangement exists.

The samples used in these studies were amorphous films, about 10 μm thick, prepared by vacuum thermal evaporation of the powdered As_xSe_{1-x} melt-quenched material at the rate of 1 μm min^{-1} onto quartz substrates held at room temperature, as well as polished mirrorlike parallelepipeds of vitreous As_xSe_{1-x}. The As_xSe_{1-x} bulk glasses were prepared according to the conventional melt-quenching method. Annealing of the films was carried out in air at ambient pressure and at temperatures below the glass transition temperature. Thin film samples were kept in complete darkness until measured to minimize exposure to light sources, which could lead to changes in the properties and structure of the films. It is important to accentuate that after the annealing procedure, their Raman spectra became indistinguishable from the corresponding spectra of melt-quenched glassy samples. It should be noted that in the Se-rich compositions studied, only slight differences in the spectra of melt-quenched and As-deposited samples were observed.

Right-angle Raman spectra were measured using a RAMANOR U-1000 spectrometer. The spectral slit width was ~1 cm^{-1}, and the excitation wavelength was 676 nm. Raman spectra of the amorphous films were recorded with sufficiently low-incident laser beam power densities, $P = 3$–5 mW, to avoid photostructural changes. The latter is known to transform the Raman spectra. The identity of the experimental spectra obtained from different points of the sample and the good reproducibility of the spectra in repeated scans show that photodarkening did not play a role in the subsequent Raman measurements. The time required to scan one spectrum in the spectral range 100–300 cm^{-1} is about 5 min.

Using the RAMANOR U-1000 spectrometer, the laser beam was focused on a 100-μm spot. Raman spectra of the amorphous films were recorded with sufficiently low-incident laser beam intensity, $I = 1.3 \times 10^2 \mathrm{W\,cm^{-2}}$, to avoid photostructural changes. The identity of the experimental spectra obtained from different points of the sample and the good reproducibility of the spectra in repeated scans show that photodarkening did not play a role in the subsequent Raman measurements.

The structure of photocrystallized films was investigated using X-ray diffraction. It is important accentuate that after the annealing procedure, their Raman spectra become indistinguishable from the corresponding spectra of melt-quenched glassy samples. It should be noted that in the Se-rich compositions studied, only slight differences in the spectra of melt-quenched and As-deposited samples were observed.

In Figure 2.7, a typical Raman spectrum of a-Se measured at a low-incident-radiation power density, 3 mW, is shown. The stable level of the scattered light intensity and the good reproducibility of the spectrum in the repetitive cycles clearly indicate the absence of any structural changes in a-Se induced by laser irradiation of such power density. On the high-frequency side (ω = 100–300 $\mathrm{cm^{-1}}$), the spectrum contains an intensive peak at 255 $\mathrm{cm^{-1}}$ and some peculiarity (shoulder) at 237 $\mathrm{cm^{-1}}$.

The features of the figure agree with previously reported data [39–43]. In the low-frequency region, one can observe the broad peak with ω_{max} = 16–20 $\mathrm{cm^{-1}}$. This so-called boson peak occurs in the low-frequency region of the Raman spectra of all amorphous and vitreous solids.

When we tried to identify certain vibration bands observed in the Raman spectrum of a-Se, some difficulties arose. Initially, the a-Se Raman spectrum was to be interpreted by analogy with sulfur, using a molecular approach. The main vibration

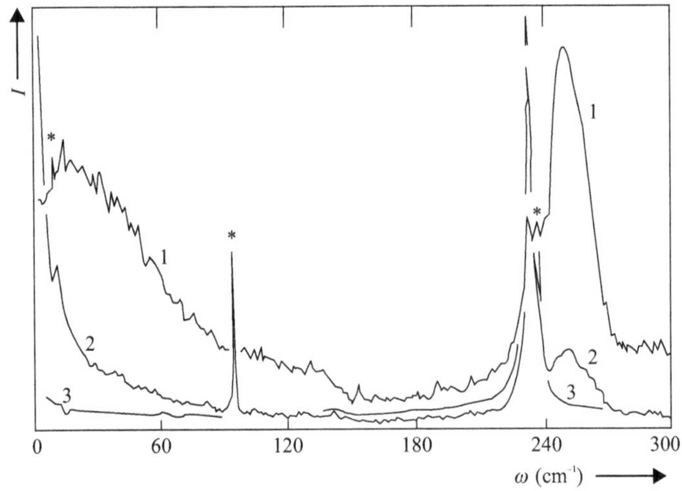

Figure 2.7 Raman spectra of (1) amorphous and (2), (3) photocrystallized selenium. Experimental details for (2) and (3) are: (2) after 5 min exposure of the sample to 10 mW, (3) after 30 min exposure to 10 mW. Asterisks indicate laser plasma lines.

band was considered to be the superposition of the peaks at 237 and 255 cm^{-1}, which were characteristic of chains and rings, respectively [41]. However, further experimental data has cast some doubts on this approach. The expectation was that a discernible difference would be observed in the contributions of 237 and 255 cm^{-1} modes to the main vibration band in the samples prepared at different conditions (e.g., substrate temperature during deposition for amorphous films or quenching rate for glassy samples). This difference would be caused by changes in rings to chains ratio. Therefore, it was clear that the spectral region 200–300 cm^{-1} was unsuitable for ring diagnostics. This was consistent with the conclusions of [42] (Figure 2.8).

The Se_8 peak (112 cm^{-1}) could not be detected in the experiment. Probably, this was connected with its weakness and indicated an evidently low concentration of rings. Consequently, the spectral features in the main vibration band, the 255 cm^{-1} peak and the shoulder at 237 cm^{-1}, were mainly associated with the chain vibrations (Figure 2.9).

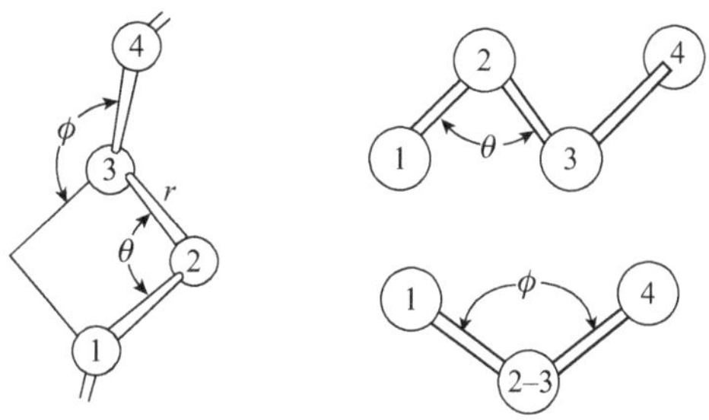

Figure 2.8 Se chain molecules and the dihedral angle φ. This involves an angle between planes and thus four atoms labeled 1, 2, 3, and 4. It is observed looking down the bond joining atoms 2 and 3.

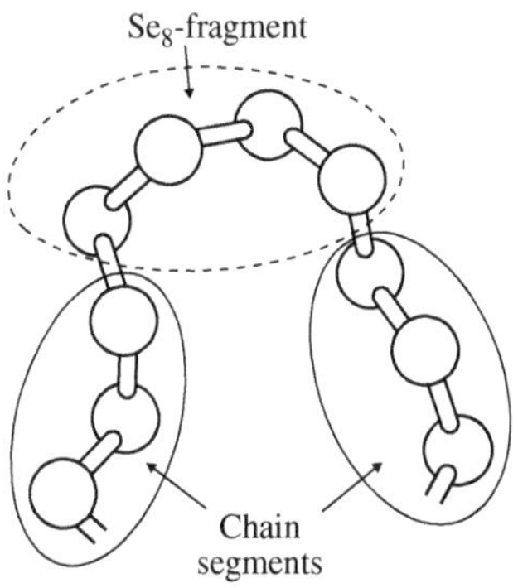

Figure 2.9 Local molecular order in a selenium chain in which there are segments characterized by repetition of the same dihedral angle (chainlike in the case of trigonal Se) and segments characterized by an alternating dihedral angle (ringlike in the case of Se_8 molecule).

2.4 Composition Dependence of Raman Bands in Amorphous Se-rich Alloys As_xSe_{100-x}

Here, the experimental results on Raman scattering spectra for Se-rich amorphous semiconductors As_xSe_{1-x} are discussed. In Figure 2.10, typical Raman spectra of amorphous Se and As–Se alloys with As content up to 5 at.% are shown. The major spectral feature in the high-frequency region is the 255 cm^{-1} band. Another prominent spectral feature, which is not shown in this figure, is the broad peak at ω_{max} = 16/20 cm^{-1}. This is characteristic of Raman scattering for all amorphous solids and glasses.

Next, we consider only the high-frequency region. The weak feature observed at 112 cm^{-1} in the a-Se spectrum diminishes with As addition and at 5 at.%, it completely disappears (see inset Figure 2.10). At the same time, the difference in the spectra in the region of the main vibration band is obvious. Thus, with increase of As content, the transformation of the Raman spectrum in this region is retraced (Figure 2.11).

The most important points are the following:

- Spectrum broadening with increasing As addition;
- Growth of scattered light intensity from the low-frequency side of the main maximum (255 cm^{-1});
- Appearance of a broadened band at 220/230 cm^{-1} (this band is the most intense in the Raman spectrum of $As_{0.4}Se_{0.6}$). It should be noted also that the main maximum is slightly shifted to higher frequency for amorphous As_xSe_{1-x} to Se (255 cm^{-1}).

The intensity of the 220/230 cm^{-1} band in the As concentration interval 0–5 at.% remains practically unchanged. Then at 6 at.% As, an increase of the band intensity occurs. A gradual intensity rise is observed for the band at 220 cm^{-1} as the As content is further increased. For As content exceeding 35 at.%, the band dominates in the Raman spectrum.

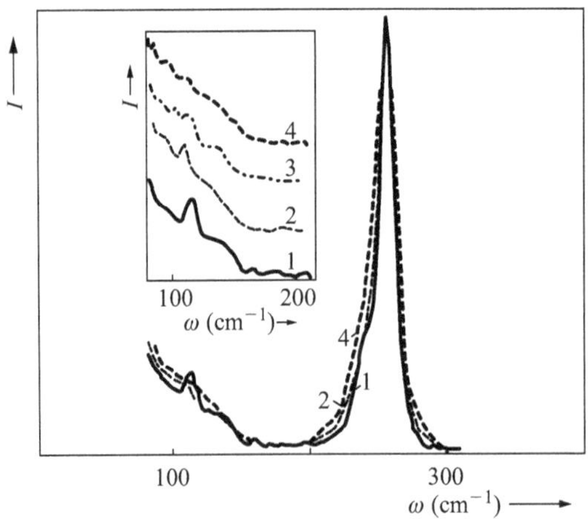

Figure 2.10 Comparison of the Raman spectra of (1) amorphous Se and As_xSe_{1-x}, (2) $x = 0.02$, (3) 0.04, and (4) 0.05 at.%. Each trace has been normalized to the same peak (255 cm^{-1}) intensity. The inset shows the bending mode region.

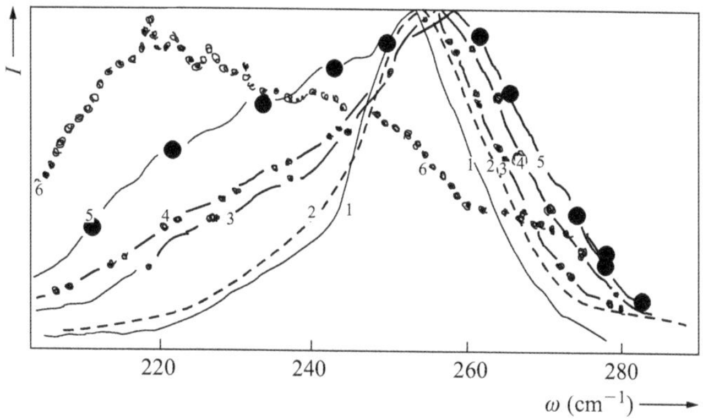

Figure 2.11 Raman spectra of amorphous As_xSe_{1-x} films annealed at $T = Tg$. (1) $x = 0$ (solid line), (2) 0.05 (dashed line), (3) 0.10 (dashed-dotted line), (4) 0.12 (dashed-double dotted line), (5) 0.20 (solid circles), and (6) 0.40 (points). We have used these unconventional denotations for Raman spectra to make them distinguishable.

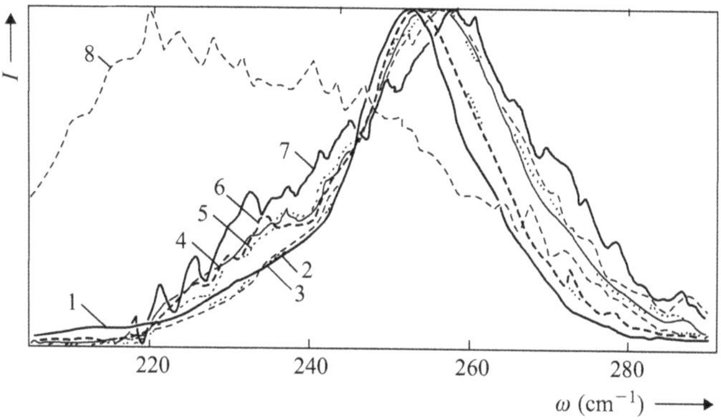

Figure 2.12 The difference spectra (see text for details) for a-As_xSe_{1-x}. $x = 0.04$, 0.05, 0.06, 0.08, 0.10, and 0.20 at.%, curves 2–7, respectively. For the sake of comparison, Raman spectra (1) of a-Se and (8) $As_{0.4}Se_{0.6}$ are also shown.

It seems to be reasonable to approximate the observed Raman spectra of As_xSe_{1-x} as the superposition of the spectra of amorphous Se and $As_{0.4}Se_{0.6}$. The corresponding calculations have been performed. These calculations yield that a systematic discrepancy between approximated and experimental spectra is observed. As for the latter, the greater values of the main peak width are typical.

Figure 2.12 shows difference spectra obtained by subtracting the $As_{0.4}Se_{0.6}$ spectrum from the experimental Raman spectra. The relative contribution of the $As_{0.4}Se_{0.6}$ spectrum was fitted to the $\sim$230 cm^{-1} region where the contribution from pure Se was negligibly small. It is obvious that after such a procedure, some peak remains, but its width and position differs from that of a-Se. Based on the data in

Figure 2.12, values of the parameter A were estimated. This parameter represents the ratio of the integrated Raman intensity in the interval limited by the typical frequencies of $AsSe_{3/2}$ unit vibrations (205–230 cm^{-1}) to the integrated intensity of the whole spectrum of valence vibrations:

$$A = \int_{205\,cm^{-1}}^{230\,cm^{-1}} I \Bigg/ \int_{205\,cm^{-1}}^{290\,cm^{-1}} I$$

(here, I is the intensity of the corresponding Raman band). Figure 2.13 shows that the dependence $A\,f(x)$ is non-monotonous—parameter A increases around ~6 at.% As. For the frequency range 240–270 cm^{-1}, the change of scattered intensity with composition has a smoother character. On the same figure, for the corresponding spectra, the dependence of the peak frequency, ω_{max}, and its width, $\Delta\omega_{max}$, on As content is displayed. It is important to note the similarity of the composition dependence of A, ω_{max}, and $\Delta\omega_{max}$.

An attempt to simulate As_xSe_{1-x} Raman spectra by a superposition of two constant spectral forms failed—one of the forms belonged to a-Se, the other to $As_{0.4}Se_{0.6}$. Onari et al. [44,45] were the first to use such an approach. The experimental spectra could be approximated assuming a considerable broadening of chain vibrations and their frequency displacements. We consider that such an approach is correct and that the difference spectra themselves are convincing arguments in favor of this: the change of the Raman spectra with composition together with parameters A, ω_{max}, and $\Delta\omega_{max}$ (see the corresponding figures) support this suggestion.

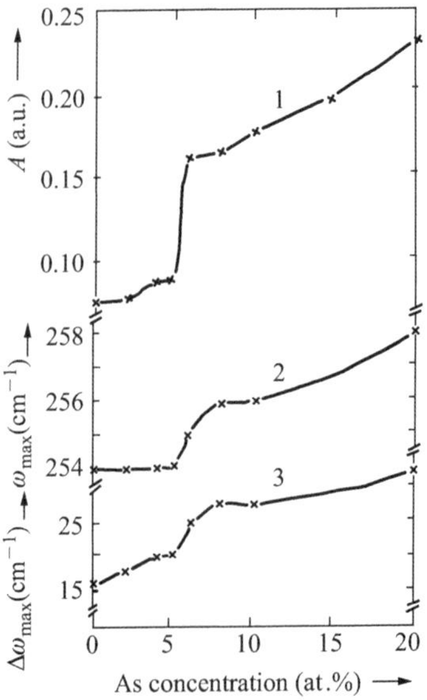

Figure 2.13 Composition dependence of the parameters (1) A, (2) ω_{max}, and (3) $\Delta\omega_{max}$.

Composition-dependent studies on the physical properties of binary and ternary chalcogenide glasses give evidence for the existence of mechanical and chemical thresholds at certain compositions of these materials [46–48]. The As–Se system displays main extrema of various properties at the stoichiometric composition (the mechanical and chemical thresholds coincide at $x = 0.40$). There seems to be an additional threshold at $0.06 \leq x \leq 0.12$ (see Wagner and Kasap [49] and present data). It can be argued that the non-monotonic behavior observed in the concentration dependence of glass transition temperature, density, etc., in this range [50] originates from changes in bond topology [46,49]. We assume that in Se-rich glasses, the network is dominated by Se atom chains (quasi-one-dimensional network), and the addition of As atoms leads to branching due to threefold coordination of As atoms. Recent publications [49,51,52] offer a new approach to the problem of local bonding in amorphous chalcogenides. The anomalous behavior near $x \approx 0.06$ is ascribed to the disappearance of Se_8-like segments. Based on this configuration, it is suggested that the number of cis-configurations in rings starts to decrease so that the mediate-range correlation is modified. The considerable reduction in the vibration mode at $\sim 112\,cm^{-1}$ associated with cis-segments in ring components strongly supports this suggestion.

Changes in the Raman spectrum with composition allow us to conclude that incorporation of As leads to cross-links between chainlike and ringlike segments of amorphous Se.

There are strong indications that the compositional dependence of physical and chemical properties has no connection with chemical ordering. In fact, the binary As_xSe_{1-x} alloys exhibit extrema in compositional dependence of the density, not only at the $As_{0.4}Se_{0.6}$ composition but also for the non-stoichiometric chalcogen-rich $As_{0.06}Se_{0.94}$ and the pnictogen-rich $As_{0.6}Se_{0.4}$ samples. This means that the x-dependence of the density originates from changes in bonding topology.

We may summarize the general features of the observed transformation of Raman spectra in the range of bond bands for amorphous selenium. (1) There is a certain threshold energy, E_{th}, of the incident radiation (Figure 2.14). (2) Below E_{th}, no changes to the Raman spectrum are observed. Here, we note that the only exception is the so-called boson peak detected at around $17\,cm^{-1}$, which is weakened by illumination. (3) Above E_{th}, an increase of the incident energy density modifies the spectra.

The present experimental results distinguish three successive stages of photocrystallization in a-Se with regard to irradiation energy density. First of all, it is necessary to point out the absence of any significant structural transformations in films and bulk samples at $E_{th} \leq 4\,J\,cm^{-2}$. This is strongly supported by the identity of the Raman spectra recorded in repetitive cycles. At the first stage, which is induced by irradiation with incident energy density $\geq 3.8\,J\,cm^{-2}$, microcrystallite formation probably takes place. In such a case, the $255\,cm^{-1}$ peaks dominate the Raman spectra.

The second stage of photocrystallization ($\sim 11\,J\,cm^{-2}$) is characterized by an enlargement of microcrystalline units. This is demonstrated by the growth of the $237\,cm^{-1}$ peak. Finally, at $\sim 20\,J\,cm^{-2}$, photocrystallization practically takes place. This stage is marked by a dramatic increase and narrowing of the peak at $237\,cm^{-1}$ with irradiation time. At the same time, the $255\,cm^{-1}$ peak becomes more and more

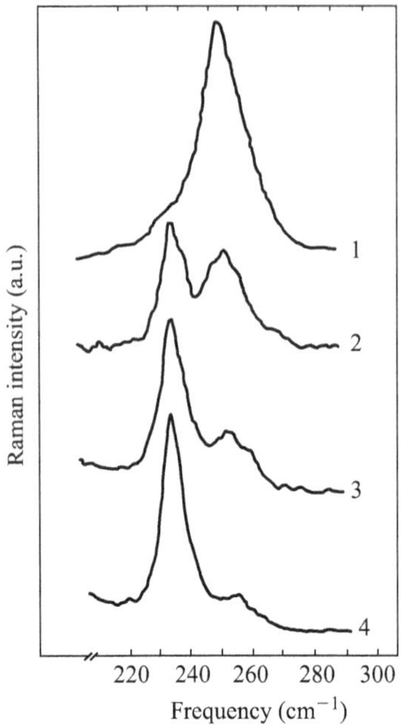

Figure 2.14 Laser-induced transformation of Raman spectra of amorphous Se. Experimental details for (1)–(4) are: (1) reference spectrum of amorphous state, (2), (3) and (4) after exposure to $E = 15$, 20 and $25 \times 10^3\,\mathrm{J\,cm^{-2}}$, respectively. $I = 1.3 \times 10^2\,\mathrm{W\,cm^{-2}}$.

suppressed with respect to other Raman active modes and, finally, it completely disappears. At this last stage of photocrystallization, the absence of the low-frequency ($17\,\mathrm{cm^{-1}}$) peak is also characteristic.

It seems to be reasonable to assume a thermal mechanism, caused by laser heating of the sample, for the observed structural transformation. This suggestion is strongly confirmed by the clearly manifested threshold behavior. Additional support comes from the fact that at low temperature, the threshold power (e.g., $20\,\mathrm{J\,cm^{-2}}$ at $T = 100\,\mathrm{K}$ for a-Se), which is necessary for changes in Raman spectra to be observable, is several times more than changes at $T = 300\,\mathrm{K}$. Note that the changes under examination qualitatively differ from the well-known photodarkening phenomena. This takes place at any value of irradiation power; threshold behavior is not characteristic of photodarkening phenomena. The magnitude of photodarkening depends mainly on the amount of absorbed energy and significantly increases with lowering temperature.

The relatively greater efficiency in films in comparison with bulk samples is an established feature of photodarkening. In contrast, in our case, the threshold power densities for a-Se films are found to be higher than for bulk samples. According to results [53,54], discernible, reversible photodarkening in a-Se at $T = 100\,\mathrm{K}$ occurs at photon energy $h\nu \geq 2.0\,\mathrm{eV}$ with efficiency maximum at $\sim 2.4\,\mathrm{eV}$. The exciting irradiation energy, $h\nu = 1.84\,\mathrm{eV}$ is probably low to induce significant photodarkening at $T = 100\,\mathrm{K}$. At the same time, the probability of transient photodarkening effects relaxing cannot be definitively ruled out after finishing the irradiation at $T = 300\,\mathrm{K}$ for $E \approx 3.8\,\mathrm{J\,cm^{-2}}$.

Table 2.1 E_{th} Values as a Function of As Concentration in As_xSe_{1-x} Alloys

As concentration (at.%)	0	5	8	12	20
E_{th}*($kJ\,cm^{-2}$)	2 ± 1	12 ± 3	7 ± 2	8 ± 2	45 ± 5

* The E_{th} values are shown for the case $I = 1.3 \times 10^2\,W\,cm^{-2}$.

With regard to As_xSe_{1-x} Raman data, principal results are the following:

- The spectra of the As_xSe_{1-x} amorphous alloy samples before irradiation were the same as those reported in [15,31,40,42–45].
- The value of E_{th} necessary for changes in the Raman spectra to be observable varies with the addition of As (Table 2.1).
- Under irradiation with $E > E_{th}$, the recorded spectra clearly show a narrowing and increase in the intensity of the $237\,cm^{-1}$ Raman band.
- In As_xSe_{1-x} samples with $x \leq 0.05$, no additional photoinduced changes (compared to those recorded for pure Se) in their Raman spectra are observed (see Figure 2.15 and compare with the results for a-Se). It should be emphasized that introducing such a relatively large quantity of As additives adds no appreciable influence on the photocrystallization product. It is well known that Se is likely to be photocrystallized at ~350 K [22,50]. Reasonably, X-ray diffraction patterns of the samples have been measured. Since the illumination region is ~3 mm in radius, the pattern is noisy. However, we can clearly see four crystalline peaks located at $2\theta = 24°$, 30°, 41°, and 45°. The peaks can be indexed, respectively, as 100, 101, 110, and 111 of the trigonal (hexagonal) Se crystal [55].
- For As content >15 at.%, the main result of this study is the appearance and disappearance of new Raman bands typical for $As_{0.4}Se_{0.6}$.

Figure 2.15 shows the appearance of an additional peak ($\sim 264\,cm^{-1}$) superimposed on the amorphous peak ($\sim 255\,cm^{-1}$). With a further increase of the irradiation energy density, crystallization starts immediately. Here, we should note an evolution qualitatively similar to that shown for pure Se and $As_{0.05}Se_{0.95}$ in the shape of spectra over the concentration range 0–20 at.%. In the final stage of photocrystallization, the spectra of As_xSe_{1-x} alloys are free of key crystalline features that occur [56] in the spectra of crystalline As_2Se_3. It is of particular significance that only the $237\,cm^{-1}$ band of trigonal selenium contributes to the spectra of photocrystallized As_xSe_{1-x} ($0 < x < 20$) films.

We define E_{th} (e.g., $E_{th} = 2 \times 10^3\,J\,cm^{-2}$ at $I = 1.3 \times 10^2\,W\,cm^{-2}$ in the case of pure Se) as the energy density for which the system is not yet disturbed structurally (on the scale of short-range order) by laser irradiation. The absence of any significant bonding changes in films is supported by the identity of the Raman spectra recorded in repetitive cycles. This result holds for both pure amorphous selenium and Se-rich As_xSe_{1-x} alloys. For energies less than E_{th}, reversible photodarkening and transient transmission changes are observed. These effects are characteristic of the amorphous phase, and the system remains in the amorphous phase under irradiation and even afterward. The lack of any noticeable variation in the transient behavior T_{rel} for samples of different substrate materials and film thicknesses excludes the

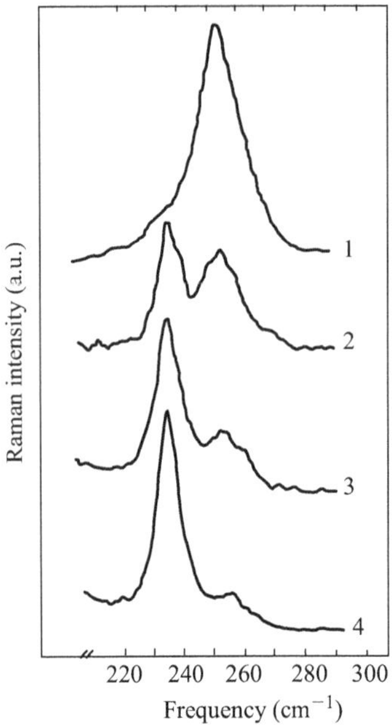

Figure 2.15 Development of the photocrystallization effect in $As_{0.05}Se_{0.95}$ as a function of exposure: (1) reference spectrum of amorphous state; (2), (3) and (4) after exposure to 15 and 25 $\times$ $10^3\,J\,cm^{-2}$, respectively. $I = 1.3 \times 10^2\,W\,cm^{-2}$.

possibility that the the effect is due to small changes in sample temperature during and after illumination, or to photoinduced heating.

We have recently reported similar dynamic photoinduced changes in some photoelectronic properties detected by TOF and xerographic techniques [18–20,57]. These experiments may provide the first evidence that deep defects can be altered temporarily by room-temperature irradiation. Note that there is a close correlation between the recovery of optical parameters and photoelectronic characteristics in exposed samples. Although a complete correlation of microscopic structural modifications with macroscopic photodarkening phenomena must await further experimental measurements, it is only natural that we relate the transient changes in the transmission with changes in deep defect states. We identify such centers as arising from native (thermodynamic) structural defects (e.g., C_3^+ and C_1^- in amorphous selenium). Band-gap light can probably initiate conversion of traps of small cross-sections to those of larger cross-sections [8,58].

Above E_{th}, all the observed irreversible changes may be attributed to optical constant variations and modifications in the kinetics of light-induced crystallization. The previously mentioned experimental results resolve successive stages of photocrystallization in a-Se. In such a photocrystallization process, amorphous Se undergoes a transformation to trigonal selenium, which is the most stable modification. Raman scattering studies together with X-ray diffraction data give an unambiguous indication of trigonal selenium.

On the basis of the present Raman data, we conclude that the features of the photocrystallization effects in As_xSe_{1-x} alloys with $x < 0.15$ are qualitatively the same as those in amorphous selenium. Some deviation in precrystallization behavior of $As_{0.2}Se_{0.8}$, namely, the appearance and disappearance of the weak quasi-crystalline peak at 264 cm^{-1}, probably indicates As_2Se_3-like cluster creation and annihilation. This could be clusters with a more ordered structure than that existing in the amorphous phase. At the same time, they are not yet microcrystallites with inherent Raman peaks. It is reasonable that the environment of the clusters prevents their growth and transformation into microcrystallites. Other Se clusters reach the critical size required for microcrystallite formation. After that, sample exposure to $E > E_{th}$ crystallized selenium, while the As-containing clusters remained in the amorphous phase. This explanation agrees with the results of the study of laser-induced structural transformations in glassy As_2Se_3 [58] and also with the mechanism proposed by Phillips [59].

It is known that As is an effective additive to decrease the tendency toward crystallization [50]. Our experimental results, namely, the greater value of E_{th} for a-As_xSe_{1-x} films, indicate that the addition of As affects the suppression of the crystal nucleation and growth in amorphous selenium. The long Se chains in amorphous selenium branch at the site of As atoms. The length of Se chains becomes short [45,60], and the amorphous Se cannot easily crystallize with an increase of As concentration.

At the same time, it is necessary to note that the changes in optical transmission and diffraction efficiency that occur are not monotonic with increasing As content. It seems to be reasonable to connect such behavior with some discontinuity of atomic arrangements with increasing As content. Our recent study of composition dependence of Raman bands in amorphous As_xSe_{1-x} supports this suggestion.

2.5 Raman Scattering in Pure and Alloyed Amorphous Selenium: Low-Frequency Spectral Region

It has become evident that traditional techniques of structural investigation are inefficient for the study of disordered materials. They give information on the structure of a short-range order only and are slightly sensitive to more long-range correlations in arrangement of atoms. The existence of the medium-range order is admitted: periodic atomic arrangements inherent in crystals remain within several coordination spheres and then are violated. It is assumed that the character of the violation may depend on preparation and chemical composition of the sample [28,61]. A number of experimental procedures were proposed for measurement of a size-ordered microregion, or structural correlation range R_c. One of these methods determines R_c by the position of a low-frequency boson peak in the Raman spectra [62].

It has been shown that the Raman intensity can be described by

$$I/\varpi(n(\varpi)+1) = \sum_B c_b(\varpi) g_b(\varpi)/\varpi^2,$$

where $g_b(\varpi)$ is the density of states and $c_b(\varpi)$ is the matrix element of the coupling of the vibration mode b to the light. $n(\varpi)$ is the boson occupation number. In the low-frequency region, $I/\varpi(n(\varpi) + 1) \sim \varpi^2$ increases and then reaches a maximum at a certain value $E = E_{max}$ (Figure 2.16), which depends on chemical composition and thermal prehistory of the sample [63,64]. This peak is called a boson peak. It accounts for 30–90% of integrated intensity of Raman spectrum in glasses.

Malinovsky et al. [61] have compared the spectral form of the low-frequency band in various samples. For the sake of simplicity, all the spectra are given in the same energy scale $E_n = E/E_{max}$ (Figure 2.17). A unique property of the boson peak was found: in a broad-frequency range, the form of the peak is independent of the chemical composition of the sample. Additional analysis of the results (Figure 2.17) indicates that the boson peak has a universal nature, which is due to the most general features of the vitreous structure and is not associated with the concrete chemical composition of glass or the structure of corresponding crystal. One of these features is the medium-range order. Universal form of the boson peak is assumed to be due to this general character of vitreous solids.

The low-frequency region $0 < \omega < 100\,cm^{-1}$, in which the boson peak appears in a-Se, is of special interest. It has been found that the spectral form of the boson peak is nearly the same for a wide series of oxides, chalcogenides, and low-molecular-organic glasses. The universal form of the low-frequency peak

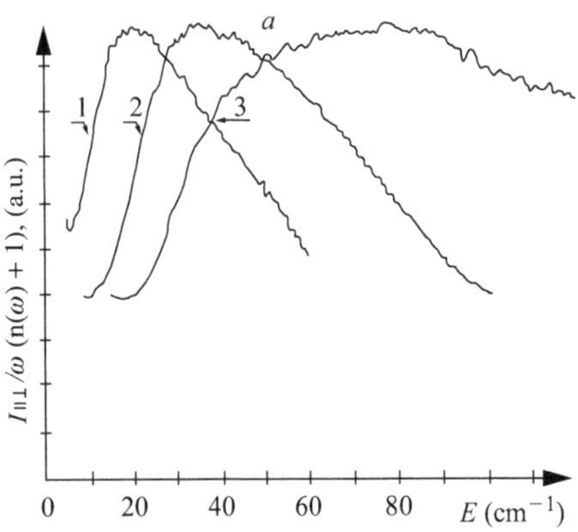

Figure 2.16 Low-frequency Raman spectra of As_2S_3—1, $Bi_4Si_3O_{12}$—2, $La_2S_3Ga_2S_3$—3 glasses [61].

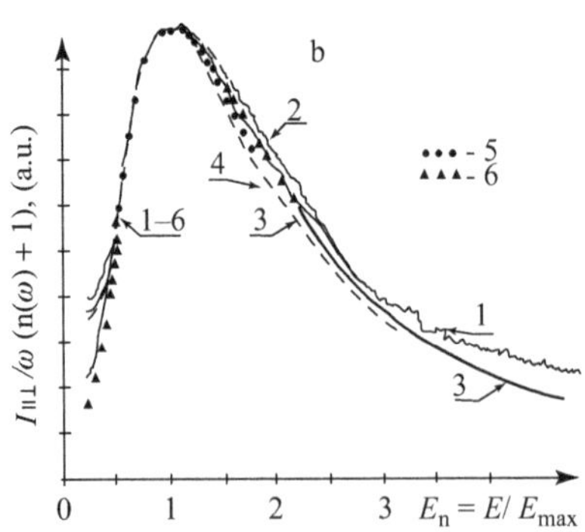

Figure 2.17 Raman spectra of different glasses in a scale $E_n = E/E_{max}$: 1—As_2S_3 ($E_{max} = 26\,cm^{-1}$), 2—$Bi_4Si_3O_{12}$ ($34\,cm^{-1}$), 3—SiO_2 ($52\,cm^{-1}$). In addition, it shows spectra: 4—B_2O_3 ($28\,cm^{-1}$), 5—B_2O_3O ($45\,cm^{-1}$), 6—GeS_2 ($22\,cm^{-1}$), and 7—Li_2O ($88\,cm^{-1}$) [61].

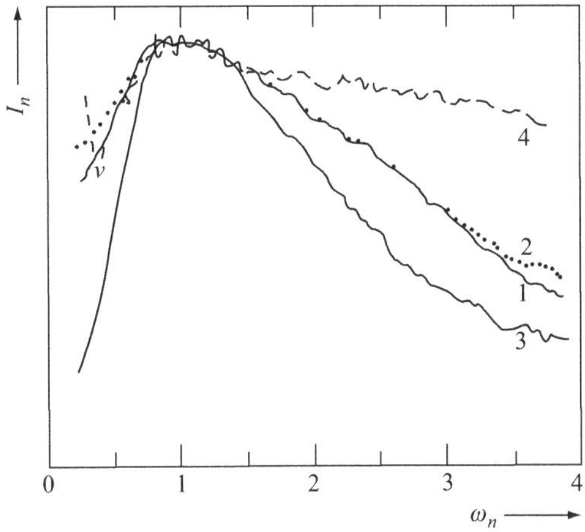

Figure 2.18 Low-frequency Raman spectra of different glasses in a scale $\omega_n = \omega/\omega_{max}$. (1) Se ($\omega_{max} = 17\,cm^{-1}$ at $T = 100\,K$), (2) Se ($\omega_{max} = 17\,cm^{-1}$ at $T = 300\,K$), (3) As_2Se_3 ($\omega_{max} = 26\,cm^{-1}$, $T = 300\,K$), (4) polymethylmetacrylate (PMMA) ($\omega_{max} = 17\,cm^{-1}$).

is due to the universality of glassy material in the scale of medium-range order $L \sim v/\omega_{max} = 1$–$2\,nm$ (v is the sound velocity). In the case of a-Se, it is observed (Figure 2.18) that the spectral form of the boson peak essentially differs from that of the majority of inorganic glasses. The spectral form of the boson peak in a-Se seems to be in between that of polymeric and low-molecular glasses.

Normalized Raman spectra $I_n = I/\omega(n(\omega) + 1)$ for a-Se and a series of other composition samples are given in the same energy scale $\omega_n = \omega/\omega_{max}$, where $n(\omega) + 1 = 1/(\exp(h\omega/kT) - 1) + 1$ is the boson factor for the Stokes component.

This result can be explained by a preferentially chainlike structure of a-Se. This may form a structure similar to that of linear polymers PMMA. With regard to its structure on the scale of medium-range order, a-Se may be placed between three-dimensional network glasses and polymeric ones. Examples of mediate-range order in elemental and compound materials, including a-Se, have been extensively discussed in [65–70].

There is another possible explanation. Se has its glass transition temperature near room temperature; therefore, it is in a well-annealed state. It is known [61] that the intensity of the boson peak relative to the main bond modes in the Raman spectrum significantly decrease as the structural order of the sample increases (e.g., in equilibrated or annealed samples). This decrease, in turn, may lead to an increased contribution from other modes to the high-frequency side ($\omega \geq \omega_{max}$) of the boson peak. This is clear if we compare the shape of corresponding peaks in a-Se and As_2S_3 (Figure 2.18).

2.6 Conclusion

The molecular structure of amorphous selenium differs from that of the majority of inorganic glasses on a scale of medium-range order. Se chains form a structure similar to the structure of linear polymers on the scale $\sim 1\,nm$.

Depending on the irradiation energy density, two qualitatively different regions are observed. Below the energy density threshold, E_{th}, only small changes in the local structure of the system can be detected.

Above E_{th}, the changes were attributed to crystallization transformation. In addition, we have detected the successive phases in such a transition, which is a threshold phenomenon.

It has been shown that Raman scattering spectra of amorphous As_xSe_{1-x} alloys change non-monotonically with composition in the region of bond stretching modes. Certain extrema in various physical properties exist at composition range 6–12 at.% As. The presence of this topological threshold is established by direct evidence, such as peculiarities in the compositional dependence of the Raman vibration modes of glasses. These peculiarities are caused by the transition from a chain-ringlike structure to a chainlike structure with an increasing degree of cross-linking.

Based on Raman scattering studies, we have shown how the structure transforms chronologically in amorphous As_xSe_{1-x} recording media.

Laser-induced changes at room temperature involve two phenomena essentially different in their origin: transient reversible changes (photodarkening) and irreversible changes (photocrystallization) with gross structural reorganization. For high values of energy density, the Raman spectrum has pronounced crystallization-related changes.

Our explanation is based on the assumption that the radiation pumps the material from an amorphous state toward a crystalline state through the formation of small clusters, which coalesce to form large clusters, attaining microcrystallite size at high-energy density levels.

2.7 References

1. S.O. Kasap, J.B. Frey, G. Belev, O. Tousignant, H. Mani, L. Laperriere et al., Phys. Status Solidi B 246 (2009) 1794.
2. M. Popescu, J. Non-Cryst. Solids 352 (2006) 887.
3. A. Madan and M.P. Shaw, The Physics and Applications of Amorphous Semiconductors, Academic Press, Boston, MA, 1988.
4. A. Feltz, Amorphous Inorganic Materials and Glasses, VCH, Weinheim, Germany, 1993.
5. V.I. Mikla and V.V. Mikla, J. Mater. Sci. Mater. Electron. 20 (2009) 1095.
6. V.I. Mikla and V.V..Mikla, J. Optoelectron. Adv. Mater. 10 (2008) 131.
7. V.I. Mikla and V.V. Mikla, Optoelectron. Adv. Mater. Rapid Comm. 1 (2007) 272.
8. S.O. Kasap, in Handbook of Imaging Materials (eds. A.S. Diamond and D.S. Weiss) (Marcel Dekker, Inc., New York, NY, 2nd ed., 2002) 329 and references therein.
9. K. Tanaka, in: Encyclopedia of Materials (Elsevier Science Ltd., 2001) 1123.
10. K. Shimakawa, Adv. Phys. 44 (1995) 474.
11. S.O. Kasap and J.A. Rowlands, J. Mater. Sci. Mater. Electron. 11 (2000) 179.
12. Z. Borisova, Glassy Semiconductors, Plenum Press, New York, NY, 1981.
13. N.F. Mott and E.A. Davis, Electronic Processes in Non-Crystalline Materials, Oxford University Press, Oxford, 2nd ed., 1979.
14. K. Hulls and P.W. McMillan, J. Non-Cryst. Solids 15 (1974) 357.
15. J. Schottmiller, M. Tabak, G. Lucovsky, and A. Ward, J. Non-Cryst. Solids 4 (1970) 80.
16. A.E. Owen, A.P. Firth, and P.J.S. Ewen, Philos. Mag. B 52 (1985) 347.
17. A.E. Owen and W.E. Spear, Phys. Chem. Glasses 17 (1976) 174.
18. V.I. Mikla, D.G. Semak, A.V. Mateleshko, and A.R. Levkulich, Sov. Phys. Semicond. 21 (1987) 266.

19. V.I. Mikla, D.G. Semak, A.V. Mateleshko, and A.R. Levkulich, Sov. Phys. Semicond. 23 (1989) 80.
20. V.I. Mikla, J. Phys. Condens. Matter 9 (1997) 9209.
21. K. Tanaka, Rev. Solid State Sci. 2 & 3 (1990) 644.
22. J. Dresner and G.B. Strinfellow, J. Phys. Chem. Solids 29 (1968) 303.
23. J.P. De Neufville, in Optical Properties of Solids—New Developments (ed., D.O. Seraphin) (North-Holland, Amsterdam, 1975) 437.
24. H. Fritzsche, in Insulating and Semiconducting Glasses (ed., P. Boolchand) (World Scientific, Singapore, 2000) Chapter 10.
25. V.I. Mikla and I.P. Mikhalko, J. Non-Crystall. Solids 180 (1995) 236.
26. M. Chomat, D. Lezal, J. Gregore, and I. Srb, J. Non-Crystall. Solids 20 (1976) 427.
27. V.I. Mikla, J. Phys. Condens. Matter 9 (1997) 9209.
28. V.P. Zakharov and V.S. Gerasimenko, Structural Peculiarities of Semiconductors in Amorphous State (Naukova Dumka, Kiev, 1984).
29. F.L. Galeneer, J. Non-Crystall. Solids 123 (1990) 182.
30. M. Cardona, Light Scattering in Solids (Springer-Verlag, Berlin, 1975).
31. R.J. Nemanich, G.A. Connell, T.M. Hayes, and R.A. Street, Phys. Rev. B 18 (1978) 6900.
32. O.V. Luksha, V.I. Mikla, V.P. Ivanitsky, A.V. Mateleshko, D.G. Semak, J. Non-Crystall. Solids 144 (1992) 253.
33. L. Cervinka, Czech. J. Phys. B 25 (1985) 1193.
34. S.R. Elliott, J. Non-Crystall. Solids 106 (1988) 26.
35. J.S. Lannin, Phys. Today 41 (1988) 28.
36. A.J. Leadbetter and A.J. Apling, J. Non-Crystalline Solids 21 (1976) 47.
37. A.C. Wright, R.N. Sinclair, A.J. Leadbetter, J. Non-Crystall. Solids 71 (1985) 295.
38. M.F. Daniel, A.J. Leadbetter, A.C. Wright, R.N. Sinclair, J. Non-Crystall. Solids 23 (1979) 271.
39. A.A. Baganich, V.I. Mikla, D.G. Semak, and A.P. Sokolov, Phys. Stat. Sol. (b) 166 (1991) 297.
40. M. Gorman and S.A. Solin, Solid State Commun. 18 (1976) 1401.
41. M.H. Brodsky and M. Cardona, J. Non-Crystall. Solids 31 (1978) 81.
42. P.J. Carroll and J.S. Lannin, Solid State Commun. 40 (1981) 81.
43. P.J. Carroll and J.S. Lannin, J. Non-Crystall. Solids 35/36 (1980) 1277.
44. T. Mori, S. Onari, and T. Arai, J. Appl. Phys. 19 (1980) 1027.
45. S. Onari, K. Matsuishi, and T. Arai, J. Non-Crystall. Solids 74 (1985) 57.
46. J.C. Phillips, J. Non-Crystall. Solids 43 (1981) 37.
47. M.F. Thorpe, J. Non-Crystall. Solids 57 (1983) 355.
48. K. Tanaka, Phys. Rev. B 39 (1989) 1270.
49. T. Wagner, and S.O. Kasap, Phil. Mag. B 74 (1996) 667.
50. Z. Borisova, Glassy Semiconductors (Plenum Press, New York, NY, 1981).
51. P. Boolchand, M. Jin, D.I. Novita, and S. Chakravarty, J. Raman Spectrosc. 38 (2007) 660.
52. E. Ahn, G.A. Williams, and P.C. Taylor, Phys. Rev. B 74 (2006) 174206.
53. K. Tanaka and N. Odajima, Solid State Commun. 43 (1982) 961.
54. R.T. Phillips, J. Non-Crystall. Solids 70 (1985) 359.
55. Powder Diffraction File (ASTM card) (ed., L.G. Berry) Joint Committee on Powder Diffraction Standard, Philadelphia, PA (1974).).
56. R. Zallen, M.L. Slade, and A.T. Ward, Phys. Rev. B3 (1971) 4257.
57. M. Abkowitz, and R.C. Enck, Phys. Rev. B27 (1983) 7402.
58. V.I. Mikla., Doctor of Sciences ThesisInstitute of Solid State Physics, Academy of Sciences, Kiev(1998).

59. J.C. Phillips, Solid State Phys. 10 (1982) 165.
60. V.I. Mikla, D.G. Semak, A.V. Mateleshko, and A.A. Baganich, Phys. Stat. Solidi (a) 117 (1990) 241.
61. V.K. Malinovsky, V.P. Sokolov, Solid State Commun. 57 (1986) 757.
62. A.J. Martin, W. Brenig, Phys. Status Solidi B 63 (1974) 163.
63. R.J. Nemanich, Phys. Rev. B 16 (1977) 1655.
64. H. Kawamura, F. Fukumasu, Y. Hamada, Solid State Commun. 43 (1982) 229.
65. G. Lucovsky and F.L. Galeneer, J. Non-Crystall. Solids 35/36 (1980) 1209.
66. J. Jakle, in Amorphous Solids: Low-Temperature Properties (ed., W.A. Phillips) (Springer-Verlag, Berlin, 1981) 135.
67. V.K. Malinovsky, V.N. Novikov, and V.P. Sokolov, Phys. I Khim. Stekla 15 (1989) 331.
68. G. Lucovsky, J. Non-Crystall. Solids 97/98 (1987) 155.
69. H. Richter, Z.P. Wang, and L. Ley, Solid State Commun. 39 (1981) 625.
70. V.A. Bagrynskii, V.K. Malinovsky, V.N. Novikov, L.M. Puschaeva, and A.P. Sokolov, Fiz. Tverd. Tela 30 (1988) 2360.

3 Effect of Thermal Evaporation Conditions on Structure and Structural Changes in Amorphous Arsenic Sulfides

Molecular structure of sulfur-based amorphous chalcogenides is significantly more sensitive to preparation conditions and various treatments compared to Se-rich materials. Vacuum-deposited amorphous chalcogenide films from an As–S system prepared by thermal evaporation methods with different rates are studied. The changes in the Raman spectra of thermally evaporated amorphous As_2S_3 films after light treatment are interpreted in terms of rearrangement of bonding configurations of molecular species that exist just after evaporation. As the evaporation temperature or deposition rate is increased, irradiation with band-gap light, As-enriching and/or even more annealing of the films induces the formation of microcrystallites. At the same time, the reversible photoinduced changes in annealed films involve negligible small changes in bonding statistics. Discernable change occurs in the medium-range order.

3.1 Influence of Preparation Conditions

An amorphous solid is a solid in which there is no long-range order in the positions of the atoms. Solids in which there is long-range atomic order are called crystalline. Most classes of solid materials can be prepared in an amorphous form. For instance, chalcogenide glasses are amorphous materials.

In principle, given a sufficiently high cooling rate, any liquid (melt) can be made into an amorphous solid. If the cooling rate is faster than the rate at which molecules can organize into a more thermodynamically favorable crystalline state, then an amorphous solid will be formed. In contrast, if molecules have sufficient time to organize into a structure with two- or three-dimensional order, then a crystalline solid will be formed.

Some materials, such as metals, are difficult to prepare in an amorphous state. Unless a material has a high-melting temperature or a low-crystallization energy, cooling must be done extremely rapidly. As a cooling is performed, the material changes from a super-cooled liquid, with properties one would expect from a liquid

Amorphous Chalcogenides. DOI: 10.1016/B978-0-12-388429-9.00003-0 state to a solid material. The temperature at which this transition occurs is called the glass transition temperature.

Amorphous solids, such as chalcogenide glasses, are rigid materials whose structures lack crystalline periodicity. Thus, the pattern of its constituent atoms or molecules do not repeat periodically in three dimensions. In the present terminology, amorphous and noncrystalline are synonymous. A solid is distinguished from its other amorphous counterparts—liquids and gases—by its viscosity.

The structural, optical, and photophysical properties of chalcogenide glasses have been the permanent subjects of systematic interest for more than 30 years [1–4]. The interest has been stimulated both by unresolved fundamental scientific problems (e.g., the structure and electronic properties of noncrystalline materials) and by the need to assess the potential of materials as optical memory elements, for image formation and other useful applications.

The structural and physical properties of amorphous chalcogenides depend on preparation conditions and subsequent sample treatments, although in a less sensitive manner than their crystalline counterparts. In many cases, the initial structures of amorphous films differ from those of well-annealed films or their bulk glasses, and they undergo significant irreversible structural changes in response to band-gap illumination or annealing (so-called photo- and thermo-induced changes). Although the induced effects have been reviewed many times, their origins remain disputable.

Based on our own experience, we examine the influence of preparation conditions (e.g., thermal evaporation rates) and of various treatments (irradiation or annealing) on the structure of fresh (As-evaporated) As_xS_{1-x} amorphous films. These films are particularly suitable objects for such a study because the irreversible changes in their optical properties, especially in As_2S_3, are much more pronounced than those in other chalcogenide glasses, such as As_xS_{1-x}.

3.2 Samples and Technique to Probe Local Structure

The samples used in these studies were usually amorphous films, typically 5–20 μm thick. They were prepared in a conventional manner, by thermal vacuum evaporation of glassy As_xS_{1-x} alloys (high-purity elements were alloyed by melting in evacuated and sealed quartz ampoules) from a molybdenum boat onto glass and quartz substrates. The evaporation was performed under two different conditions: evaporation onto a substrate with deposition rate ~20 Å s^{-1} (conventional mode) and evaporation onto a substrate with deposition rate ~1,000 Å s^{-1} (flash evaporation). The deposition rate was varied by changing the source (evaporation boat) temperature while the substrate temperature was held constant at $T = 300$ K. Chemical compositions of the films, for example, As_2S_3, were found by electron microprobe analysis to be $As_{39.7}S_{40.3}$. The lack of crystallinity in the film samples was systematically verified by X-ray diffraction measurements.

Raman spectra were measured by conventional and Fourier Transform Infrared (FTIR) Raman spectrometry (Bruker, model IFS 55). Laser irradiation of the wavelength of 647.1 and 1,064 nm from Kr-ion and Nd-YAG lasers was used for the excitation of the Raman spectra. Such low energy, especially in the latter case, is well below the values of the optical band-gaps for the glassy alloys under study. Thus, no detectable photostructural transformation took place under data collection. It is quite natural that the amplitudes of some Raman lines from As_xS_{1-x} films were often low (compared to that in bulk glass), and the spectra were usually noisy, or not smooth. The back-scattering method was exploited with the resolution of $1\,cm^{-1}$. Raman spectra of the amorphous films were recorded at sufficiently low-incident laser beam power density (3–5 mW) to avoid possible photostructural changes in the case of a 647 nm excitation line. A Kr^+ ion laser was used as an exposure source, and the films were exposed at an intensity of $30\,mW \times cm^{-2}$. Although structural transformations in As_xS_{1-x} films were the subject of several articles [5–17], some important aspects of them are still not fully understood.

3.3 Local Structure of As_2S_3 Amorphous Films

First of all, we consider the results obtained for conventionally prepared amorphous films. Figure 3.1 shows the Raman spectra for the As-deposited (1), exposed (2), annealed film (3), and bulk glass (4) forms of As_2S_3.

Qualitatively similar results for irreversible thermostructural transformation were first reported by Solin and Papatheodorou [5], Nemanich et al. [6], Mikla et al. [7], and Frumar et al. [8,9]. The spectrum of the As-deposited film consists of relatively sharp peaks on the background, typical of bulk As_2S_3 glass. These sharp features, characteristic of the As-deposited film, broaden, decrease in intensity, or even disappear irreversibly after annealing or illumination. Such a behavior

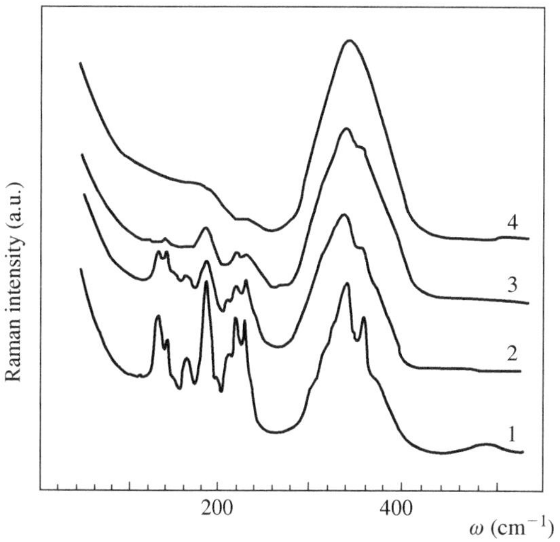

Figure 3.1 Raman spectra of As-deposited (1), photodarkened (2), and annealed (3) As_2S_3 amorphous films. For comparison, the spectrum (4) of bulk glass is also shown.

indicates that significant (in the sense of amorphous semiconductors) structural modification has taken place. Note that in bulk glass and amorphous film form with As content exceeding 40 at.%, all the relatively sharp features previously mentioned are observed. The data analysis [7–11] indicates that the sharp spectral features (except for the 490 cm^{-1} band) originate from "wrong" (like-atom or homopolar) As–As bonds in As_4S_4 molecular units. The presence of As_4S_4 molecules in the As-deposited As_2S_3 films is strongly supported by infrared (IR) [12], extended X-ray absorption fine structure (EXAFS) [13], and recent Raman scattering [7–9] measurements. Taking this into account, As-deposited films of As_2S_3 possess AsS_3 pyramidal units, which form a glassy matrix of As_2S_3 and, in addition, a partially polymerized mixture of As_4S_4 (realgar-type) and S_n molecular species.

After annealing, the spectrum of the As-deposited film becomes very similar to that of bulk glass. However, some broad features—maxima at 187 and 230 cm^{-1}—remain. This indicates that some "wrong" bonds of the type As–As and S–S exist even in well-annealed As_2S_3 films. During this treatment, the bond breaking and switching due to an increased mobility of atoms is accompanied (to a considerable extent) by structure polymerization. This is manifested in the S–As–S stretching mode (120–170 cm^{-1}), frequency spreading, and vibration band broadening typical of the bulk glass. Similar behavior in the valence mode region of As–S and As–As vibration bands takes place if we illuminate the As-deposited film. Consequently, we can say that the structure of illuminated films remains closer to that of As-deposited films.

In most studies on arsenic chalcogenide films, including those examined here, the investigators deposited amorphous layers at fairly low deposition rates ranging maximally from 10 to 100 Å s^{-1}. The reaction of these samples on band-gap illumination is manifested in an absorption edge shift to lower energies (red shift or photodarkening). In contrast, setting the evaporator at T_{evap} = 800–900°C and deposition rates of ≥300 Å s^{-1} leads to a structural modification which in turn results in a decrease in the optical band-gap. The exposed samples behave like positive photosensitive layers (in contrast to materials with negative image formation), i.e., they exhibit a blue shift of absorption edge.

In Figure 3.2, a typical Raman spectrum of a film deposited at a higher rate is shown.

In general, it is clear that on raising the evaporation temperature and deposition rate, the spectral bands sharpen and the smooth background disappears. The high-frequency component at 490 cm^{-1}, which represent S–S bonds polymerized in the glass network, disappears. This is paralleled by the appearance of additional features at 120, 150, 200, 233, and 273 cm^{-1}. Note that the strongest peak (273 cm^{-1}) in the Raman spectrum of the film deposited at high deposition rate is absent in those of As_2S_3 crystals and α, β-As_4S_4 polymorphs. Recall that this line was observed [14] in the Raman spectrum of the sample of crystalline As_4S_3 obtained by vacuum sublimation of melted As_4S_3.

As the deposition rate increased, the films were found to become sulfur-deficient. This was the case even for sulfur-rich compositions (Figure 3.3).

One of the reasons for the As-enrichment of the samples may be the partial fragmentation of As_2S_3 into As_2S_2 and sulfur during the deposition. Note that the increase in quenching temperature has a similar effect on the structure of

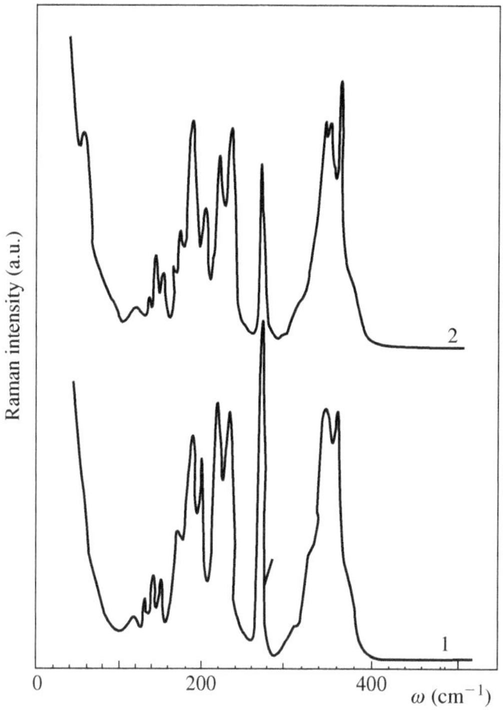

Figure 3.2 Fast-deposited As_2S_3 film Raman spectra. Spectrum (1) was taken before annealing and spectrum (2) after annealing.

melt-quenched As_2S_3, as was demonstrated by Young et al. [13], Mikla et al. [7], and Zitkovsky and Bolchand [15].

Only a slight change in the Raman spectrum was observed after irradiation of such films. In contrast, annealing resulted in further narrowing of the bands, splitting, and intensity redistribution. The low-frequency region of the annealed films exhibited a distinct shoulder at $60\,cm^{-1}$, which probably corresponds to the lattice (intermolecular) line in the spectra of the crystals α,β-As_4S_4 [11].

It is important to point out here that the transformation of the Raman spectrum with increasing As content (above stoichiometric $As_{0.4}S_{0.6}$ composition) is similar to that observed for As_2S_3 with the increase in deposition rate (for comparison, see Figures 3.2 and 3.4). Results suggest that the high deposition rate induces the development of the microcrystallite formation. Annealing further enhances the crystallization processes, and the films become polycrystalline. Such a radical structural transformation, observed in the films deposited at high rates, was found to be irreversible—the initial structure (and the corresponding absorption edge position) could not be restored even by annealing near the glass transition temperature, T_g. Moreover, the films deposited at higher rates always appeared smooth but took on a dusty appearance after light treatment and especially after heat treatment.

Returning to the films prepared in conventional mode (slow-deposited films), we next examine how photodarkening of well-annealed films and bulk samples

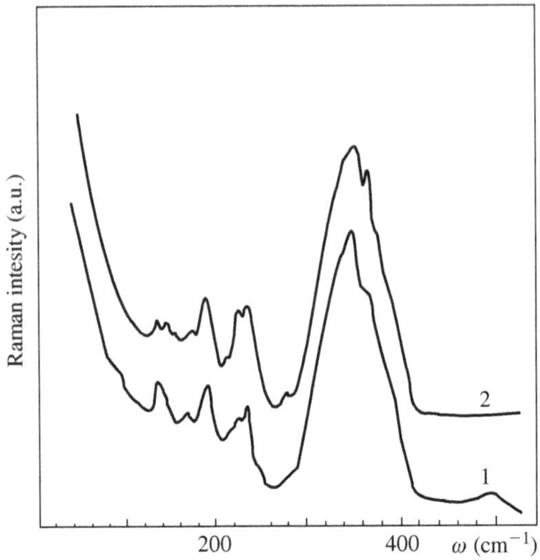

Figure 3.3 $As_{0.30}S_{0.70}$ films Raman spectra. Curve (1) is the spectrum of the film prepared in a conventional manner (slow-deposited), curve (2) corresponds to the fast-deposited film.

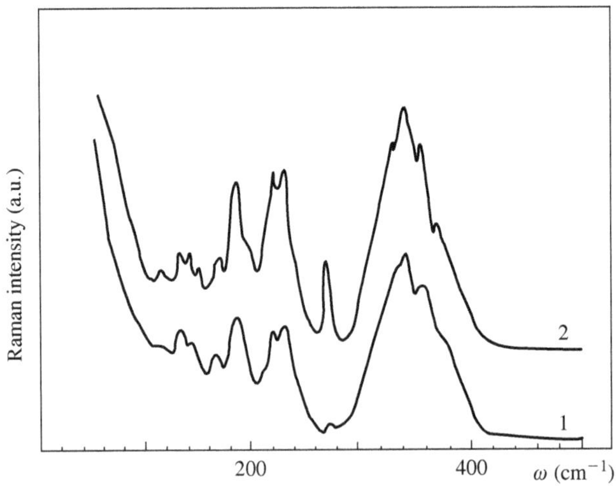

Figure 3.4 Raman spectra of $As_{0.44}S_{0.56}$ films prepared by slow (1) and fast (2) evaporation.

influences their structure. In this case, the irreversible changes were much less developed, principally because the magnitude of the reversible photodarkening is small. Structural changes detected by means of direct structural probes (e.g., X-ray diffraction [16]) were subtle. Regrettably, indirect probes of local structure by means of Raman spectroscopy also show only negligibly small changes between annealed and irradiated amorphous films. This is even in the case with the use of Fourier

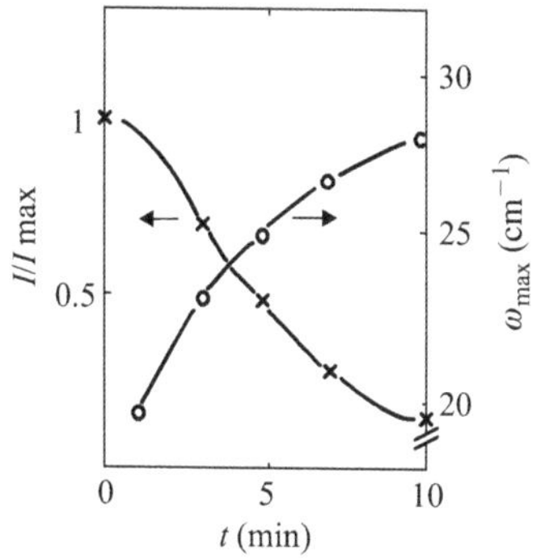

Figure 3.5 The dependence of the low-frequency peak position and relative intensity on exposure time. I_{max} denotes the peak intensity for a non-darkened sample.

Transform (FT) Raman spectroscopy, which is believed to be more sensitive than conventional Raman spectroscopy. At the same time, it should be noted that Raman scattering is sensitive to mediate-range order and can provide some structural information in terms of correlation length. In Frumar et al. [8], an increase in the intensity of the 231 cm^{-1} Raman band characteristic of As–As bonds in a-As_2S_3 with photodarkening was reported. In contrast, in the present experiments, any slight change in Raman spectra in the frequency range 100–500 cm^{-1} could not be detected. Only low-frequency Raman spectra show discernible changes under illumination.

Figure 3.5 is a summary of the low-frequency Raman spectra measurements at room temperature. As_2S_3 glassy samples are illuminated for different times from 1 to 240 min. It may be observed that the peak position of the spectrum gradually shifts to higher frequencies as the illumination time is increased. The reversible photodarkening in As_2S_3-annealed thin films and bulk glasses is also accompanied by changes in the intensity of the low-frequency peak. It is clear that the photodarkening has appreciably reduced the intensity of so-called boson peak.

The irreversible structural changes that occur in the conventionally prepared (slow-deposited) thin films upon optical illumination can be understood in terms of the mechanism described in Refs. [5–12,17]. These changes are associated with a photopolymerization of As_4S_4 and S_n molecular species. The resultant structure after such transformation is two-dimensional, layerlike, and with nearly restored chemical ordering.

The rapid deposition, as observed in the Raman data, significantly enhances compositional as well as structural disorder in amorphous samples compared with that of the stoichiometric composition. An increase in As concentration leads to a condensation of As_4S_3 and some other As-rich molecular species; a disordered molecular solid is obtained. These As-rich compositions are unstable, and presumably phase separation occurs upon thermal annealing and light irradiation. Without going further into the details of these irreversible effects in fast-deposited films, it may be suggested that irreversible photoinduced structural changes in conventionally prepared amorphous films are quite different from a phenomenological point of view and with respect to their origin.

On the basis of present Raman results and in contrast to the irreversible structural changes, one can conclude that any changes in local structure appear to accompany the reversible photoinduced changes observed in well-annealed amorphous films. This means that covalent (intra-molecular) bond breaking is unlikely to be the dominant mechanism responsible for reversible photoinduced effects in amorphous chalcogenides.

The most significant photostructural change observed using the Raman scattering technique was in the region of the low-frequency or boson peak at the wavelength of $25\,cm^{-1}$. This Raman peak signifies a certain medium-range order in the amorphous structure [18–20]. The position and intensity of the peak can be used to estimate the structural correlation length $R_c \sim (\omega_{max}/V)^{-1}$, where ω is the frequency of the peak maximum and V is the sound velocity. It can also be used to characterize the degree of structural ordering. The structural correlation length is generally associated with the dimensions of some clusterlike structure. For a-As_2S_3, the R_c value is ~7.6 Å. The clusters may exist in the form of linked pyramids. In spite of the absence of definite conclusions on the nature of low-energy excitations in disordered solids, it is obvious that differences between the boson peak in the annealed and photodarkened states are due to changes in medium-range order. A decrease in the peak height and a shift to higher ω values may be attributed to an increase in structural randomness. This is consistent with an increase in structural disorder under illumination detected by EXAFS spectra [13,21]. In spite of our experimental results on Raman scattering, photodarkening is a photostructural change and is the result of changes in medium-range order (i.e., weakly linked AsS_3 pyramids move with respect to each other, as proposed in Ref. [22]). This mechanism does not involve the breaking of covalent bonds. At the same time, some authors [22–26] suggest modification of short-range order at reversible photoinduced changes, on the basis of data obtained by indirect probes of local atomic structures. Clearly, unique diffraction-like experiments are needed; they are probably the only experiments that can give unambiguous evidence of reversible photoinduced changes in short- or medium-range order. The Raman data presented in this chapter shows that the main problems of structural change remain unresolved.

3.4 Conclusion

The influence of thermal evaporation conditions and subsequent light and heat treatments on the atomic structure was studied. The major experimental findings of these studies may be summarized as follows.

The pronounced irreversible changes observed in the Raman spectra of slowly deposited films with illumination are direct evidence for gross structural changes caused by optical irradiation. Such behavior has been interpreted in terms of a photopolymerization mechanism.

The films were found to become sulfur-deficient as the deposition (evaporation) rate increased, because of partial fragmentation of As_2S_3 into As_3S_2 and sulfur during the deposition. Fast-deposited films are unstable and phase separation occurred upon light irradiation.

The reversible photoinduced effects in well-annealed amorphous films involve structural modifications only in the medium-range order.

3.5 References

1. A. Madan and M.P. Shaw, The Physics and Applications of Amorphous Semiconductors (Academic Press, Boston, MA, 1988).
2. A. Feltz, Amorphous Inorganic Materials and Glasses (VCH, Weinheim, 1993).
3. S.O. Kasap, in Handbook of Imaging Materials, (eds., A.S. Diamond and D.S. Weiss) (Marcel Dekker, Inc., New York, NY, 2002) 329 and references therein.
4. K. Tanaka, in Encyclopedia of Materials (Kidlington, UK, 2001) 1123.
5. S.A. Solin and G.N. Papatheodorou, Phys. Rev. B 15 (1977) 2084.
6. R.J. Nemanich, G.A.N. Connell, T.N. Hayes, and R.A. Street, Phys. Rev. B 12 (1978) 6900.
7. V.I. Mikla, Y.M. Vysochanskij, A.A. Kikineshi, et al., Sov. Phys. J. 11 (1983) 73.
8. M. Frumar, A.P. Firth, and P.J.S. Owen, Philos. Mag. B 50 (1984) 453.
9. M. Frumar, Z. Polak, and Z. Cernosek, J. Non-Crystall. Solids 256 & 257 (1999) 105.
10. M.L. Slade and R. Zallen, Solid State Commun. 30 (1975) 387.
11. P.J.S. Even, M.L. Sik, and A.E. Owen, Solid State Commun. 33 (1980) 1067.
12. U. Strom and T.P. Martin, Solid State Commun. 29 (1979) 527.
13. S.Y. Young, D.E. Sayers, and M.A. Paesler, Phys. Rev. B 36 (1987) 8122.
14. A. Bertoluzza, C. Fagnano, P. Monti, and A. Semerano, J. Non-Crystall. Solids 29 (1978) 49.
15. I. Zitkovsky and P. Bolchand, J. Non-Crystall. Solids 114 (1989) 70.
16. K. Tanaka, Appl. Phys. Lett. 26 (1975) 243.
17. A.E. Owen, A.P. Firth, and P.J.S. Ewen, Philos. Mag. B 52 (1985) 347.
18. A.J. Martin and W. Brenig, Phys. Status Solidi B 63 (1974) 163.
19. R.J. Nemanich, Phys. Rev. B 16 (1977) 1655.
20. V.K. Malinovsky and A.P. Sokolov, Solid State Commun. 57 (1986) 757.
21. L. Gladen, S.R. Elliott, and G.N. Greaves, J. Non-Crystall. Solids 106 (1988) 189.
22. G. Pfeiffer, M.A. Paesler, and S.C. Agarwal, J. Non-Crystall. Solids 130 (1991) 111.
23. G. Lucovsky, M. Popescu(eds.) Non-Crystalline Materials for Optoelectronics, (INOE Publishers, Bucharest, 2004).
24. K. Tanaka, in Handbook on Advanced Electronic and Photonic Materials and Devices (Nalwa, San Diego, CA, 2001).
25. V.M. Lyubin and M.L. Klebanov, Photo-Induced Metastability in Amorphous Semiconductors (VCH, Weinheim, 2003) Chapter 6.
26. M.A. Popescu, Non-Crystalline Chalcogenides (Kluwer Academic, Dordrecht, 2000).

4 The Big Invention of the Twentieth Century—Xerography

Xerography, or the *xerographic process*, was invented by *Chester Floyd Carlson* in the 1930s, and it has played a major role in the evolution of the *Information Age*. This invention provides a cheap, simple, and reliable way of reproducing and creating documents, and it has *revolutionized* the way information is processed and disseminated. An important part of this technology is the photoreceptor, which optically *captures information*. Accurate photoreceptor performance requires the properties of *photoelectronic*, *electrostatic*, *mechanical*, and *fabrication* materials, which are amorphous. An overview of the xerographic process, its basic steps, and the relevant physical phenomena involved in the formation of latent imagery are presented. Although the emphasis is on the electronic properties of amorphous selenium, some properties of other disordered solids will be mentioned in order to amplify features typical of the disordered solid state. This chapter offers interesting information about the history and market of xerography and discusses alternative powder marketing technologies. The physics within each step are provided, as well as a guide to the literature.

4.1 Introduction

Nowadays, xerography is the technology used in all commercially available copiers, and it promises to be the most prevalent printer technology in the future. This chapter has been written to assist both the newcomer and those who are already in the field to understand better this important and complicated technology.

Xerography, the technology that started the office copying revolution, was born unheralded on October 22, 1938. It was the inspiration of a single man working in his spare time [1]. When he died in 1968 at the age of 62, Carlson was a wealthy and honored man, Xerox annual revenues were approaching the billion dollar mark, and the whole world was making copies at the push of a button. Undoubtedly, in the very beginning, this invention must have looked like a miracle.

The astounding success of xerography is all the more remarkable because it was given little hope of surviving its infancy. The development of xerography has led to new technologies that some predict will eventually eradicate traditional offset printing machines. These new machines that print in full color (such as Xeikon) use xerography but nearly provide the quality of traditional ink prints.

Amorphous Chalcogenides. DOI: 10.1016/B978-0-12-388429-9.00004-2

Today, xerography is a foundation stone of a gigantic worldwide copying industry, including Xerox and other corporations which make and market copiers and duplicators that produce billions and billions of copies a year.

The xerographic photoreceptors over the last decade have been using more organic photoconductors rather than selenium alloys; this trend is expected to continue. At the same time, some large-volume copying applications still use a-Se alloys because they provide many copies per drum. There is another challenge to the chalcogenide photoreceptors. This comes from a-Si:H photoreceptors—they have good sensitivity in the red and IR regions and exceptionally long machine lifetimes.

The following sections are for readers who may be new to xerography. Several books and review articles are now available which discuss the practical aspects of electrophotography, and general literature in this area is growing. Electrophotographic printing, as Carlson called it, was later to be named xerography after the Greek words *xeros* and *graphos*, which together mean "dry writing." An extensive bibliography on this topic is provided at the end of this chapter [2–16]. We begin with an introduction of xerography and a discussion of its technical history and the current status of the market. We can trace the evolution of subsystems beginning from Carlson's first concepts in 1937 to present-day technology. The market of xerography really began with the introduction of the first automatic copier by the Haloid (now Xerox) Corporation in 1959. Since then, the copier business has evolved into a multi-billion dollar industry with many of the world's largest corporations participating. The already large xerographic printer business is expected to grow even faster in the coming decades as the demand for computer output devices continues to increase.

The only potential competitors to xerographic printing are two related powder-marking technologies, magnetography and ionography. In magnetography, magnetic forces replace the electrophotographic forces used in xerography. In ionography, the latent image is created by placing ions on a dielectric surface, eliminating the need for a photoreceptor.

4.2 History of the Big Twentieth-Century Invention and the Greatest Inventor

More than 100 years ago, in 1906, Chester Floyd Carlson, the inventor of xerography, was born. He would become a man whose genius would forever change the practice of how people share information. His invention ultimately generated a document management industry worth more than $112 billion.

During the 1930s, Carlson was searching for a simple copying device. At this time, the only copying method available was the Photostat process. This was based on silver halide photography. These copy machines were only available at a few service centers, and the copies produced were reversed. In this process, the customer gives a paper negative—white letters on a black background. The diazo process requires ammonia fumes to develop the blue illuminated diazonium compounds coated on paper. The earlier blue print process produced white lines on a blue background and was initiated by UV exposure of iron salts. As these processes were only for engineering copying techniques, Carlson clearly recognized the need for a better copying process. Several alternatives evolved during the 1940s, including Eastman Kodak's Verifax process (a process also based on silver halide photography), 3M's Thermofax process (in which a special paper is developed by heat produced by the absorption of light in the printing), and Gevaert's and Agfa's diffusion transfer process (in which the unexposed silver salts in the positive image on film diffuse to another sheet of paper where they are chemically reduced into a positive image).

The two ideas that Carlson brought together in 1937 were as follows:

1. The formation of an electrostatic latent image using photoconductivity to selectively discharge a surface-charged insulator.
2. Development of this latent image by dusting it with electrostatically charged powders.

Photoconductivity and *electrostatics* fit together remarkably well. Electrostatic charging of materials is still a little understood, highly empirical, mostly ignored aspect of solid state physics. Photoconductivity of insulators was basically an unstudied science at the time of Carlson's invention.

Carlson was familiar with prior experiments and patents in which electrostatic images were developed with charged powders. For example, he traced the development of charged powder from Lichtenberg to Selenyi. In 1777, Lichtenberg observed starlike patterns on insulators when dust settled onto a cake resin that had been sparked. In Hungary, in 1936, Selenyi demonstrated an electrographic recording system in which a charged pattern was written on an insulator by controlling a cathode current to the insulator surface with a grid (Figure 4.1) [10].

Up to now, it is not obvious just how Carlson came upon the idea to use thin photoconductor insulators for latent image formation. By 1937, he had conceived of the electrophotography process. In October 1938, with the help of Otto Kornei, an unemployed engineer, the process transformed to practical form and was filed for patent in April 1939. Somewhat latter, in 1942, Carlson obtained a patent for this invention.

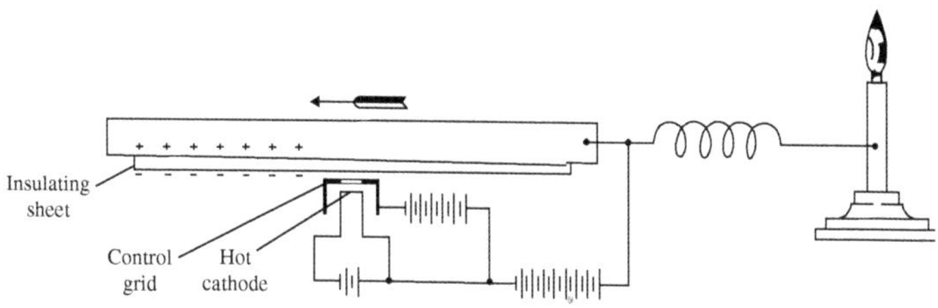

Figure 4.1 Selenyi's electrographic recording system. A charged pattern is written on an insulator by controlling a cathode current to the insulator surface with a grid. A screw arrangement permits motion of the insulator surface in a spiral path for scanning the full page. By dusting with an insulating powder, the image is developed. The candle is used to erase the latent image [10].

The first photoreceptors were composed of pure sulfur which had been fused and spread onto a metal plate and allowed to harden. Later, plates of sublimated anthracene layers with higher light sensitivity were used. The photoreceptor was charged by rubbing it vigorously with a soft material, such as a cotton or a silk handkerchief. There was an alternative method—placing a transparent conductive plate parallel to the photoreceptor. When a voltage was applied between the back of the photoreceptor and the transparent conductive plate in the presence of illumination, the top surface of the photoreceptor became charged if the light and the voltage were removed. To create the latent image, the plate was secured to the back of a camera where the image of the original was focused on the photoreceptor. Development (for creating a real image) was accomplished by sprinkling powder from a can with a cloth or fine wire screen across its mouth. Because of fusing requirements, pulverized resins were preferred. Gum copal, gum sandarac, ordinary rosin, sealing wax, dyed lycopodium powder, talcum powder, carbon dust, etc., were also used. The dusted plate was then subjected to a gentle draft of air by using the breath to blow on it. Transfer of the powder to paper was accomplished by carefully laying the paper on the photoreceptor carrying the dusted image and firmly pressing against the surface with a felt block. To improve the transfer, an adhesive such as plain water, wax, or another soft or sticky substance could be applied to the paper.

Beginning in 1940, Carlson unsuccessfully tried to enlist commercial support for his invention with companies such as RCA, Remington Rand, GE, Eastman Kodak, and IBM. Finally in 1944, Carlson entered into a royalty-sharing agreement with Battelle Memorial Institute, a nonprofit research organization in Columbia, Ohio, and joint development of the process began under Roland Schaffert. Later, John Dessauer, the director of research at a little company called Haloid in Rochester, New York, showed company president Joe Wilson an article on Carlson's electrophotography in *Radio News*. Wilson and Dessauer decided to take a gamble on the new copying technology and signed an agreement with Battelle in 1946. Haloid's funds, plus funds from the US Army, which had an interest in military photographic applications, accelerated progress in developing the technology. The process was first publicly announced and demonstrated by Battelle and Haloid at the annual meeting of the Optical Society of America in Detroit in October 1948. At that time, the term xerography came into use.

During the 10 years after Battelle became involved, many basic inventions made automatic copying a reality. Among these inventions was Bixby's discovery that amorphous selenium layers prepared by vacuum evaporation onto aluminum were photoconductive insulators with much higher light sensitivity than sulfur or anthracene. A corona-charging device was also invented at Battelle. This invention replaced Carlson's rubbing techniques. As a following step, Walkup invented the screen-controlled corona unit, which greatly lessened the danger of photoreceptor damage due to overcharging. Walkup and Wise invented cascade development, a system in which two powders, toner, and a much larger sized carrier were mixed together. By carefully choosing the surface material of the carrier and toner, most of the toner would be charged with one sign and would electrostatically adhere to the carrier. The carrier with its attached toner would then be literally cascaded over the latent image, depositing toner in the process. Another very important invention was made by Schaffert. He proposed the electrostatic transfer method, in which the back of plain paper was corona-charged and electrostatically attracted toner in the process. These inventions formed the heart of the first automatic copier introduced by Haloid (Xerox) in 1959, the Model 914. 914 was so named because it could reproduce documents up to 9 × 14 in. in size. It made seven copies in a minute and produced a revolution in the office. It is difficult to imagine an office today without a copier.

During these early years, much of the information obtained was highly empirical and not suitable for reporting in scientific journals. Status reports on the technology appeared in 1965 in the form of two books, one by Roland Schaffert entitled *Electrophotography* and one edited by John Dessauer and Harold Clark called *Xerography and Related Processes*. These books demonstrate interesting aspects of the technology.

During the 1960s, Xerox expanded its product line, introducing the 2400 series of copiers (40–60 cpm). Speed limitations and the inability of cascade development to reproduce solid areas brought about work on new development systems, including electrode cascade and later the magnetic brush systems, invented at RCA. IBM introduced its first copiers in 1970. IBM was the first to use organic photoreceptors coated onto aluminized Mylar in place of amorphous selenium. Being much softer, it had a substantially shorter life. To resolve this problem, the photoreceptor was made in the form of a long belt, which was unrolled slowly from the inside of a drum. As a result, the organic photoreceptor belt had an overall longer life than the amorphous selenium drum. By 1975, IBM was able to introduce its first computer printer based on electrophotography. The technology inside an electrophotographic printer is virtually identical to a copier with the exception of the exposure system (in place of lamps, a laser beam is scanned across the photoreceptor surface).

By the late 1970s, it was becoming increasingly clear that semiconductor lasers had significant advantages over gas lasers (they could be packaged in transistor-size containers and could be simply modulated). At the same time, there were serious disadvantages in fact that their output wavelengths were in the IR range (near 800 nm). Thus, the interest in electrophotography was reinitiated to find IR-sensitive photoreceptors.

Major developments in the technology came from Japan during the period of 1970–1980. Canon made significant new discoveries. First of all, Canon developed a new photoreceptor. It consisted of two layers—an upper insulating layer and a bottom photosensitive layer made from calcium sulfide. A latent image was produced by simultaneously exposing and charging, causing the charge of the latent image to reside at the interface between the two layers.

We present only a brief retrospective glance at electrophotography. It seems clear now that there are always new ideas to be considered with this technology, even 70 years after its invention.

Carlson invented electrophotography to take care of the need for a simple, inexpensive device that would allow office employees to copy any type of document. Carlson graduated with a BS degree in physics just as the depression was deepening. After putting out more than 80 job applications, he finally received an offer from

Bell Labs in New York. Within a year, Bell Labs had massive layoffs, and he was again out searching for a job. He landed at P. R. Mallory in New York and worked in their patent department. In this job, he realized with frustration how difficult it was to make copies of legal documents. Six carbon copies had to be made of typewritten material, and drawings had to be sent out for days or sometimes a week for photocopies. Carlson was convinced that a good office copier would be a great asset to every office. He studied evenings and weekends at the New York Public Library, gathering information about interactions of light and matter. In time, he learned of the work of Selenyi in Hungary [10]. In this work, electrostatic charges were deposited on an insulating film and then dusted with powders to make a visible image. With this in his mind, Carlson realized one day that electrostatic latent images could be created by a pattern of light projected onto a charged photoconducting film. This novel concept, Carlson said, did not come in a series of logical steps: it was simply there in a single flash of insight.

Carlson experimented in his kitchen with sulfur coatings for his photoconducting film. He melted crystals of sulfur on a zinc plate over his kitchen stove. When the sulfur sometimes caught fire, his experiments became frowned upon in the kitchen. After more than a year of experimenting, Carlson paid an unemployed engineer, Otto Kornei, to assist him. About 3 weeks later, they made *the world's first xerographic image*. It said: *10/22/38 Astoria*.

The first electrophotographic system consisted of a photoconductive layer of fused sulfur, a developer medium of lycopodium powder, and a sheet of waxed paper. The images were produced in a four-step process.

Undoubtedly, Carlson might have been discouraged had he known that *it would be another 21 years before his dream of the electrophotographic copier would be realized in the marketplace*! But he would have been heartened if he had known then that, within his lifetime, 200 million copies would be made by this process every day. (After 2000, that figure stands at more than 10 billion copies per day.)

What was said about a better mousetrap did not seem to apply to the copier; the world did not beat a path to Carlson's door. He disclosed his process to 20 of the biggest business equipment companies in America, and for 6 years he met with "*an enthusiastic lack of interest*," as he later described it.

Finally, he made contact with Battelle Memorial Institute on an unrelated project for Mallory. He told them, incidentally, about his copier invention. Their interest in metallurgy had led them to set up a graphic art department, so they agreed to explore this process.

The agreement required partial funding by Carlson, and it was more than he could afford, so his cousin Roy invested in the process, too.

Scientists at Battelle had made very substantive improvements in many parts of the process that were essential to its commercial viability. For example, they completed research on corona charging, which Carlson suggested and had already patented. The first images were made by rubbing a sulfur plate with cloth for frictional charging. Walkup and Wise invented a two-component developer that made cleaner, darker images [12]. Schaffert invented the electrostatic transfer, so the developed powder image could be transferred to ordinary paper [13].

One breakthrough invention was made by Bixby, who worked at Battelle [14]. Bixby discovered that *selenium*, when co-evaporated with sulfur, increased the sensitivity of the sulfur without destroying its ability to retain a charge in the dark. After many experiments, the optimum mixture turned out to be 0% sulfur, 100% Se. *Crystalline selenium was known to be photoconductive, but it was much too conductive to hold static charges in the dark. The glasslike vitreous selenium turned out to be highly insulating in the dark and more than 1,000 times more sensitive to light than sulfur.* It took Bixby many months of bootlegged experiments, with his wife keeping him company in his lab, evenings and weekends, to find the exact conditions to evaporate thin films of amorphous Se without the crystallization that leaked away surface charges.

Only then was the world's second type of photography a reality. Only then could the plate respond to camera lens exposures of seconds, rather than hours for each image. Only then was Carlson's dream of the office copier a realistic goal. And only then could the resources for research and development of such an ambitious undertaking be justified. Just before this point, the Haloid Company learned of the invention and negotiated a license in 1947 to commercialize the process. One young man at Haloid, Gene Fuerst, was assigned the task of demonstrating the process to the board of directors and company officers before they acquired the rights. Knowing the process was still unreliable, Fuerst wisely secreted a previously made copy within reach. When his image was developed and transferred to paper, he placed it into a poorly controlled small oven to fuse powders. When he saw smoke, he quickly retracted the copy, stamped it out on the floor to put out the flame and deftly reached for his earlier sample. His quick thinking and nimble hand may have saved the future of the copier industry.

Through the early 1950s, Haloid's research and engineering efforts could be described as probing. It was clear they had a tiger by the tail, and it would not easily be harnessed. In spite of many scattered projects exploring various applications, it is still not clear that it should have been done differently. The capital investment needed for a marketable office copier was more than Haloid could afford, and loans were available only after evidence of some commercial viability was demonstrated. A manually operated flat plate machine, the Xerox model A copier, was introduced in 1950 and upgraded to the Xerox model D processor in 1953. It took 3 min to make a copy and, fortunately, it could make paper offset masters at the same rate. This became the major xerographic product and was useful for learning more about the processes of photoreceptor preparation, charging, development, transfer, cleaning, and fusing. A microfilm enlarger (Copyflo-11) was also marketed in the 1950s, beginning in 1954.

Researchers were investigating powder cloud development for continuous tone images, X-ray images (xeroradiography), master-making processes for spirit duplicating (ditto), and stencil duplicating (mimeograph) masters. Researchers were excited to build a base of new knowledge and explore new process simplifications and new potential uses for xerography. In a short study of vapor fusing, they had fun building an S2 Schlieren optics system and seeing the flow patterns of trichloroethylene vapors. One young man working on etching resists made xerographic decals of cartoons and pictures and transferred them to surfaces everywhere. When he displayed pictures of the members of the board rather freely, his colleagues wondered if he might be jeopardizing his career.

Through all of this, Carlson worked in the labs with the others, mostly pursuing ways to make the process simpler and more affordable. He spent years transferring electrostatic latent images to dried paper, then developing it. He also showed that dried paper could be placed in contact with the latent image, and the fields reaching through the paper could be developed on the paper, eliminating the steps of transfer and cleaning. This was called interposition development. These ingenious simplifications carried their own subtle complexities, however, and never became practical. One problem is that ordinary paper is literally one million times more conducting at 85% relative humidity than at 5% relative humidity.

By 1955, the pressures were building: RCA, IBM, and Horizons were becoming active in xerographic research. Haloid, uneasy about the magnitude of costs for development, manufacturing, and marketing of copiers offered a license to IBM. Fortunately, they retained a market research organization, A. D. Little, to evaluate the market. After a thorough study, they concluded the process was too complicated to be affordable. To be fair, it should be pointed out that a latent market is not easily evaluated.

Joseph C. Wilson, CEO of Haloid, then commissioned Ernst & Ernst to conduct a similar study. Their conclusion was that whatever market existed for the office copier would saturate at about 4,000 units. Wilson must have spent many uneasy nights reflecting on the reactions of his shareholders. But he trusted his instincts and his technologists. In an act of great courage, he decided to move forward with plans.

The market predictions could hardly have been further off the mark. Instead of situating 4,000 units, about 200,000 Xerox 914 mainframes were built (some were upgraded to the higher-speed Xerox 720). About 250,000 desktop 813s were placed—including a double speed version, the Xerox 660. It was predicted that each 914 would churn out about 10,000 cpm; the average turned out to be nearly 100,000 cpm.

There was no denying that a major new industry had been born. In the past 28 years, it has grown to about $50 billion per year. Including the manufacturers, vendors, marketers, and suppliers, about 500,000 jobs have been created. And it is still growing, with extensions now into electronic printing.

The Carlson–Wilson team was a classic partnership of inventor and champion. Most inventors have neither the desire nor the ability to organize the funding, manufacturing, and marketing functions necessary for commercializing even the best of inventions. Carlson and Wilson each fully appreciated the essential strengths of the other. Both were genuine heroes, having a profound impact on the long-term industry, even beyond their own life spans. They changed the very culture of the office all over the world and improved the means of communication in the global economy.

4.3 Classification of Xerographic Processes

In 1961, an effort was made to classify xerographic (electrophotographic) processes. In this classification, xerography came to mean any imaging system in which visible or UV electromagnetic radiation was employed to produce a latent image in the form of an electrostatic charge pattern. The charge pattern in turn was assumed to produce an electric field variation (δE), which acted upon electrostatically charged pigment particles to render the latent image visible. Each pigment particle carried some constant charge Q, which served as the handle by which the electric field selectively pulled the pigment particle to the charge pattern.

After 1961, a variety of new electrophotographic imaging systems were introduced which clearly belonged to a different class than xerography. In these systems, the electromagnetic radiation acted upon the pigment to produce a latent image in the form of a charge variation (δQ) directly in the pigment particles. In other words, the process of carrying a charged pigment developer to a latent electric field image was no longer required. In such systems, the latent image was developed by applying a constant electric field (E) that sorted the pigment particles according to their charge.

A broad description of xerography was formulated by considering the total variation in electric force QE over the image plane. The total variation in QE is

$$\delta(QE) = Q\,\delta E + E\,\delta Q. \tag{4.1}$$

Xerography has two subclasses:

1. Charged pigment xerography in which the charge Q resides on pigment particles.
2. Noncharged pigment xerography in which the charge Q does not reside specifically on pigment particles (e.g., frost xerography).

The second term in Eq. (4.1) is identified with the newer class of imaging systems where the latent image is a charge variation δQ in the pigment.

4.4 Logical Steps in Practical Xerography

Xerography is a complex process (see [1–8] for other books and reviews on xerography) typically involving seven distinct process steps, which are shown schematically in Figure 4.1 and in one possible configuration in Figure 4.2. At the same time, it is necessary to note here that the number of steps in the xerographic process *depends on the type of photoreceptor that is used in the system.* It takes seven steps to produce a single copy when the photoreceptor is to be reused in a multiple-copy situation. It takes four steps when the photoreceptor is a photoconductive paper that is used to produce only one copy.

The basic steps for a multiple-copy situation are shown in Figure 4.3. Here, the seven steps of the xerographic process are indicated schematically. These steps include:

1. Charging or sensitizing the photoreceptor.
2. Exposing the photoreceptor to form the latent electrostatic image.
3. Developing the electrostatic image.
4. Transferring the image to paper.

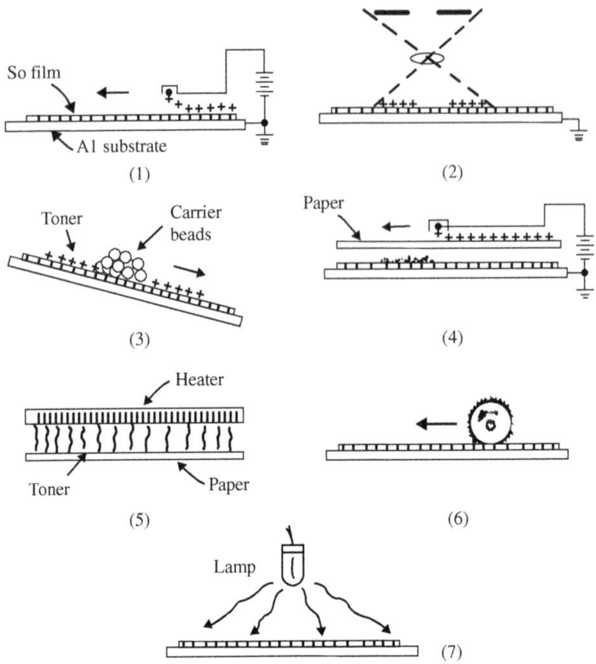

Figure 4.2 The main steps of the xerographic process, including charge, exposure, development, transfer, fuse, and clean (after [8]).

5. Fusing or fixing the image onto the paper.
6. Cleaning the extra toner from the photoreceptor.
7. Erasing the electrostatic image.

If one uses photoconductor paper, some steps, namely (4), (6), and (7), are eliminated and the image is fused onto the paper.

4.5 Realization of Xerographic Process

Here, we discuss the construction and function of an office photocopier. The photocopier has a metal cylinder mounted to rotate about a horizontal axis, which is called the drum. It is about 150 mm (6 in.) in diameter, and the end-to-end dimension is the width of the print to be produced plus a generous tolerance. The drum is manufactured by vacuum deposition with a surface coating of amorphous selenium (more recently ceramic or organic). Amorphous selenium will hold an electrostatic charge in darkness and will conduct away such a charge with the application of light.

The drum rotates at the speed of paper output. One revolution passes the drum surface through the steps described later. The drum may be a belt and there are variants at every step.

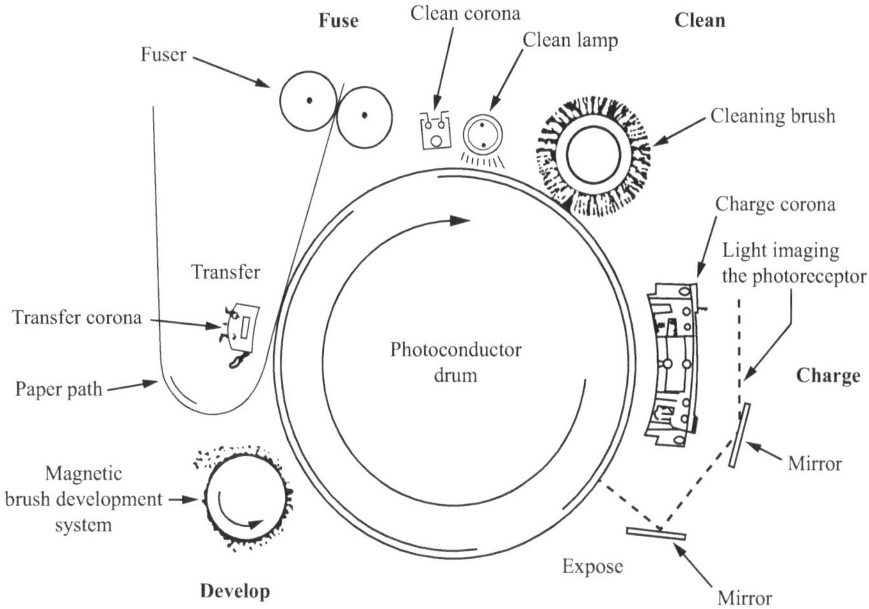

Figure 4.3 The seven basic steps of the xerographic process: (1) sensitizing or charging, (2) exposure, (3) development, (4) image transfer, (5) fixing or fusing, (6) cleaning, and (7) erasing [9].

4.6 Phenomenological Aspects

We shall discuss the phenomenological aspects of each step of the xerographic process. We begin each step with a short characterization.

4.6.1 Sensitization (Charging)

A xerographic photoreceptor is sensitized by uniform charging. In the absence of light, the photoreceptor is an insulator.

The sensitization is effective when the surface of the photoreceptor is charged uniformly to its maximum surface charge density. The charge per unit area σ_p is determined (a) by the potential V required for development (~600 V or more) and (b) the photoreceptor thickness. The photoreceptor is made as thin as possible, for charge transport reasons, but thick enough to avoid dielectric breakdown. As an example, for organic photoreceptors whose dielectric strengths are ~30 V μm^{-1}, 20 μm thicknesses are necessary. Then, applying Gauss' Law, one can obtain the required photoreceptor charge per unit area

$$V = E_p d_s = (\sigma_p / \varepsilon_0)(d_s / K_s), \quad (4.2)$$

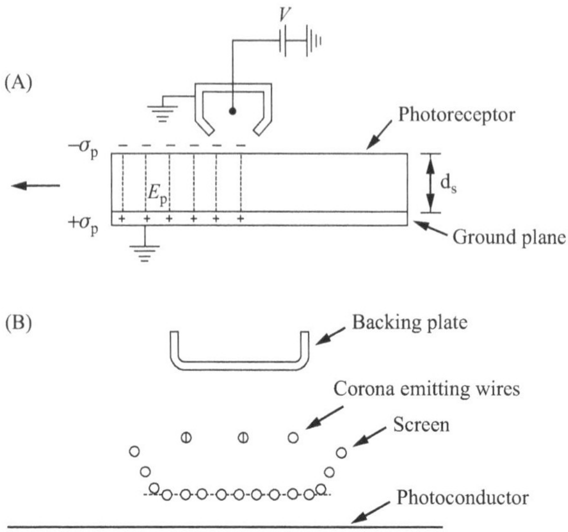

Figures 4.4 Schematic of devices used for charging a photoconductor. In (A), a corotron is shown. In (B), a scorotron is shown, and a screen is placed between the corona wires and the photoreceptor. The potential of the screen determines the approximate maximum potential to which the photoreceptor will be charged [8].

where E_p is the electric field internal to the photoreceptor, d_s is the thickness of the photoreceptor, K_s is the photoreceptor dielectric constant, and ϵ_0 is the permittivity of free space. For amorphous Se photoreceptors (with dielectric strength $K_s = 6.6$ and $E = 10\,\mathrm{V\,\mu m^{-1}}$), this gives $\sigma_p = 60\,\mathrm{nC\,cm^{-2}}$.

The most practical method of sensitization is to charge with a corona discharge and a device used to place the charge on the photoreceptor, which is called a corotron (Figure 4.4A). If a sufficiently high voltage is applied between a fine wire and a grounded shield, the air near the wire will become ionized. Ions of the same polarity as the wire will be swept by the electric field toward the photoreceptor. While the shield current is a source of inefficiency, it provides stability to the corona by forcing the operation conditions to be far from the threshold. The nature of the ionized molecules resulting from the corona is CO_3^- and $(H_2O)_nH^+$, with $n = 4$–8 for negative and positive coronas, respectively.

In principle, the voltage necessary to create the ionized air depends primarily on the wire diameter. Typically, a $50\,\mu\mathrm{m}$ diameter wire spaced 1 cm from the ground plane, operating at a potential 8,000 V, is sufficient to charge a photoreceptor that moves at a speed of $5\,\mathrm{cm\,s^{-1}}$.

The corona appears as a uniform blue-white sheath around the wire for positive polarity; for a negative wire, there are glowing bluish points spaced at regular intervals along the wire. They represent a significant source of nonuniformity for negative charging.

A corona discharge is simply a high-voltage discharge obtained by applying a several thousand volt bias between a conductive wire and the substrate of a photoreceptor plate. The extremely high fields near the conductive wire ionize the air, producing ions that drift along the field lines between the wire and the photoreceptor and deposit on the free surface of the photoreceptor plate. The amount of charge supplied to the surface depends on several factors:

- the voltage on the corona wire,
- the spacing between the wire and the photoreceptor,
- the length of time the ion current is allowed to flow.

The magnitude of the surface potential produced by the corona charge is equal to σ/C, where σ is the total charge per unit area supplied to the material and C is the capacitance per unit area.

A corotron is simply a corona wire that is surrounded by a ground shield (Figure 4.4A). The shield serves two important purposes. First, it reduces variations in the charging current produced by irregularities in the wire diameter. Second, it compensates for a nonuniform spacing between the corotron and the photoreceptor.

A scorotron is essentially a modified corotron with a control grid. The control grid, similar to a corotron, serves two purposes. First, *it further improves the uniformity of charging*, and second, *it protects the photoreceptor from overcharge*.

The control grid is biased (charged) at the desired photoreceptor surface potential. The control grid allows ions to flow to the photoreceptor surface until the surface potential is equal to the grid potential. At that potential, the current stops and the surface potential will remain at the grid potential. The corotron and scorotron are well described in Ref. [16].

The discussed methods are the most convenient, efficient, and simple. The reader may find other methods in the scientific literature, but it is necessary to note that they are not suitable for practical use. These methods include charging with ions generated by a radioactive source exposed to air, charging by rolling a biased rubberized roller over the photoreceptor surface, and induction charging by bringing a biased electrode close to a grounded photoreceptor and removing the biased electrode. In principle, any process that can put a uniform charge on the surface of a photoreceptor can be used in a xerographic system.

4.6.2 Exposure

Light, reflected from an image, discharges the normally insulating (at darkness) photoreceptor and produces a latent image—a charge pattern on the photoreceptor that mirrors the information to be transformed into a real image.

The second step in the xerographic process is the formation of the latent electrostatic image. In principle, the latent electrostatic image is the electrical charge pattern on the surface of the photoreceptor, which is produced by exposing the sensitized photoreceptor to the optical image. This is achieved in the following way: the document to be copied is lit and passed over a lens so that its image is projected onto the drum at exactly the same speed that the drum turns. The light from the input image impinges on the surface of photoreceptor. The incident photons are absorbed by the photoreceptor, creating free electron–hole pairs in proportion to the intensity variation in the optical image. The electron–hole pairs are separated by the electric field, and each carrier drifts toward the appropriate electrode: in the case of positive charging of the photoreceptor, the holes drift toward the substrate and the electrons drift toward the illuminated surface. When the carriers reach the appropriate electrode, they recombine with the surface charge. Reasonably, the number of photogenerated carriers varies with the incident light intensity. As a result, a latent electrostatic image patterned after the optical image appears on the surface of the photoreceptor.

One method of measuring the surface potential is the following: the photoreceptor is charged to its optimum potential by a corotron in one position and transported to a second position where the photoinduced discharge measurements are performed. The surface potential is monitored by a thin wire (or transparent Nesa glass) probe, which measures the magnitude of the potential induced on the probe. The surface potential is measured as a function of exposure by monitoring the surface potential while simultaneously exposing the material through a thin wire (or Nesa glass) probe. The resulting surface potential versus exposure curve is defined as the photoinduced discharge curve (PIDC). This curve is used as the transfer function which relates the broad area optical input to the surface potential and eventually to broad area development.

In a laser or LED (light emitting diode) *printer*, a thin beam is modulated by the controller. Where there is text or image on the document, light will not penetrate to the drum; the opposite is true where there is no image. The light that penetrates releases the charge on the drum, but only where there is no image. The resulting charge that remains on the drum (the latent image) is a positive of the original document.

When discussing the development of the latent electrostatic images, it is more correct to talk about surface potential rather than surface charge densities. There are two reasons for this. First, it is the surface potential, not surface charge density, that is measured experimentally. The surface charge density is determined from the potential through the expression $\sigma = CV$, where C is the capacitance of the sample per unit area and V is the measured surface potential. Second, the electric fields which drive the development processes are usually proportional to the surface potential and the differences in surface potential (contrast potentials).

Photoreceptors were originally fabricated by evaporating amorphous selenium onto metallic drums. In amorphous selenium, carriers of both polarities (holes and electrons) are mobile. *This is unique to amorphous chalcogenides, where, as a rule, the mobile carriers are exclusively holes*—even though hole mobility is of nearly two orders of magnitude greater than electron mobility (but it remains easily measurable). Usually only hole transport is used and, therefore, the top surface is positively charged.

Much of the scientific literature devoted to xerography has focused on the microscopic processes occurring inside photoreceptors. Photons of light produce electron–hole pairs near the surface of the photoconductor. The electrons and holes are then separated by an internal electric field, and the holes drift across the thickness of the photoreceptor in the presence of the electric field. Where there is text or image on the document, the corresponding area of the drum will remain unlit. Where there is no image, the drum will be illuminated and the charge will be dissipated. The charge that remains on the drum after this exposure is a latent image and is a positive of the original document. The time that charge carriers need to access the opposite (bottom) surface of the photoreceptor is small compared to the xerographic cycle time ($\sim$1 s).

4.6.3 Development

Electrostatically charged and pigmented polymer particles, called toner, $\sim$10 μm in diameter, are brought into the vicinity of latent image. Due to the electric field created by the charges on the photoreceptor, the toner adheres to the latent image and transforms it into a real image.

The latent electrostatic image could be developed by some relatively simple methods. Returning to the technical history of xerography, one of the methods consists in pouring powder over the material, as Carlson and Kornei did when producing the first xerographic prints. During this process, some of the powder particles acquire an electrostatic charge, which is opposite to that of the latent image. These particles are attracted to the surface of the photoreceptor and adhere in numbers proportional to the strength of the latent image and produce a visible image. Another method is simply to blow smoke over the latent image. Some of the particles in the smoke are electrostatically charged, and those particles of the correct polarity will be attracted to the latent image. Generally, charged pigment particles of any kind can develop the latent image provided that they are sufficient in number and of the correct polarity. During development, toner particles are attracted to the electrostatic latent image on the drum surface. Thus, a visible toner image is created.

Several xerographic systems have been invented to develop the latent image on a large-scale practical basis. These development systems include the following:

- cascade,
- magnetic brush,
- liquid ink,
- aerosol,
- electrophoretic,
- frost,
- fur brush.

With the exception of frost and liquid ink, these systems all use very fine charged pigment particles to develop the image. The differences between the charged pigment systems lie in the method of transporting the pigment particles to the latent image. Let us consider each of them.

Cascade development is one of the most common methods of xerographic development. In this system, the developer material consists of *toner and carrier* beads. The toner is a fine pigment powder which is triboelectrically attached to a much larger carrier bead. The carrier bead is primarily a transport vehicle which brings the toner to the electrostatic image. The development is accomplished by flowing or cascading the developer material over the photoreceptor plate. The agitation of the developer shakes of some of the toner, which senses the electric field produced by the image and drifts to the surface of the photoreceptor. Here, some of the toner is electrostatically pulled from the carrier beads when the beads come in close contact with the surface. This situation can occur when the electrostatic forces between the photoreceptor and the toner become larger than the adhesive forces between the toner and the carrier.

A development electrode can be used in cascade development. The development electrode is a grounded or biased conducting plate that is placed close to the surface of the photoreceptor. In this case, the developer is cascaded between the development electrode and the surface of the photoreceptor. The development electrode enhances the electric fields in the development zone, and this, in turn, enhances the development process.

The two-component systems are usually used for high-speed machines. They are called two-component because the developer mix has two components, toner and carrier. The toner particles (polymer-loaded with carbon black as a colorant) have average diameters of 10 μm and are developed onto the latent image on the photoreceptor and then transferred and fused on paper. The carrier is an approximately 200 μm diameter magnetically soft sphere coated with a polymer chosen to correctly charge the toner on contact. The purposes of the carrier beads are twofold: to charge the toner by static electrification and to transport the toner into the vicinity of the latent image. Thus, the drum is presented with a slowly turbulent mixture of toner particles and larger metallic carrier particles. The carrier particles have a coating which, during agitation, generates a triboelectric charge (a form of static electricity), which attracts a coating of toner particles. The mix is manipulated with a magnetic roller to present to the surface of the drum/belt a brush of toner. With carrier contact, each neutral toner particle has an electric charge of polarity opposite to the charge of the latent image on the drum. The charge attracts toner to form a visible image on the drum. To control the amount of toner transferred, a bias voltage is applied to the developer roller to counteract the attraction between toner and latent image.

Where a negative image is required, as when printing from a microform negative, the toner has the same polarity as the corona in the first step (charging). Electrostatic lines of force drive the toner particles away from the latent image toward the uncharged area, which is the area exposed from the negative.

Early color copiers and printers used multiple-copy cycles for each page output, using colored filters and toners. Modern units use only a single scan to four separate miniature process units, operating simultaneously, each with its own coronas, drum, and developer unit.

Magnetic brush development is an important example of a development system, and it is used today in almost all copiers and printers operating above 30 cpm. The system consists of a magnet, a mass of iron beads or filings, and toner particles. These toner particles are electrostatically attached to the iron beads. The mass of iron filings is attracted to the magnet, forming long chains that look like bristles on a brush, hence the name magnetic brush. Development is accomplished by passing the latent electrostatic image beneath the magnetic brush. The electric field produced by the image electrostatically strips the toner from the iron beads and deposits it onto the surface of the photoreceptor.

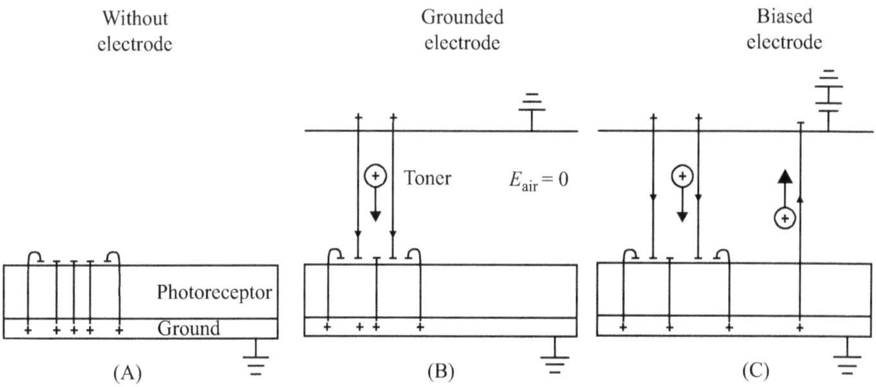

Figure 4.5 (A) External electric field associated with a latent image only appears in the form of fringe fields around the edges of charges in the absence of an electrode. (B) In the presence of an electrode, some electric field is capacitively coupled to the electrode. (C) A bias on the electrode creates a gray scale control and can be used to lower background development.

The electric field in the magnetic brush developer can be larger than that in electrode cascade development. This is because there is greater bead density and magnetic alignment of the beads (this produces a smaller effective spacing between the magnetic brush and the photoreceptor surface).

The latent image, consisting of surface charge patterns on the photoreceptor, produces electric field lines connecting the surface charges to image changes, which reside in the ground plane of the photoreceptor in a configuration without an electrode (Figure 4.5). In Figure 4.5A, we note only that the charged area, which extends above the photoreceptor in distance, is comparable to the thickness of the photoreceptor. Obviously, close approach of grounded counter electrodes (Figure 4.5B), such as in the case of a roller, draw field lines outward in proportion to the ratio of the capacitance per unit area of the photoreceptor and the counter electrode. The charged toner introduced into this region is attracted by the electric field to the photoreceptor. Note that the electrode allows additional functions (Figure 4.5C). A bias on the electrode can create a force that draws the toner away from the background regions, and it can decrease the electric field above solid area latent images. Additional data can be found in Refs. [7,8].

Electrophoretic development of the image takes place with liquid developers. These developers consist of a suspension of charged pigment particles in a dielectric liquid (e.g., insulating hydrocarbon). The development proceeds by covering the latent image with the developer fluid. The charged particles in the fluid, because of the electric field produced by the image, drift along these field lines to the surface of the photoreceptor.

Fur brush development has similarity with the magnetic brush development system. The toner is triboelectrically attracted to a cylindrical brush from animal fur. The toner is continuously supplied to the fur brush, which rotates and oscillates while in contrast with the photoreceptor. It is important to note that humidity changes have a great effect on the triboelectric properties of the fur.

Frost development is a process where the image is observed as a deformation in a resin layer (and not as the density of pigment particles on the surface of the photoreceptor). In this type of development, a transparent insulating resin film is placed over the surface of the photoreceptor. As a next step, the resin–photoreceptor layer is charged and simultaneously exposed to the input image. This produces large fields in the exposed resin layer. The magnitude of these fields varies in proportion to the total amount of exposure. After that, the resin layer is softened (by means of heat or solvent vapor). This procedure allows the resin layer to deform in proportion to the field across the photoconductor film. The development image appears as a frostlike image due to the light-scattering effect on the deformation layer. One can use the image permanently or reuse it by heating the layer to relax the image.

4.6.4 Image Transfer

The developed toner on the photoconductor is transferred to paper by corona charging the back of the paper with a charge opposite to that of the toner particles.

To produce a xerographic copy, it is necessary to transfer the developed toner image from the photoreceptor to the paper. The more practical methods of transferring images are as follows:

- *Electrical transfer*: This is carried out by placing paper over the image and charging the paper with a corotron of the same polarity used to charge (sensitize) photoreceptor. The charging produces a strong electrostatic field in the space between the paper and the toner on the surface of the photoreceptor. This field lifts the toner from the photoreceptor and transfers it to the paper, producing an image on the paper. As described in the following steps, the paper is removed from the photoreceptor and the image is ready to be fixed (fused) permanently.
- *Adhesive transfer*: The image on the photoreceptor can also be transferred by pressure-sensitive adhesives coated onto the paper. In this case, the adhesive force between the paper and the toner exceeds the electrostatic force between the toner and the photoreceptor. The transfer is made by placing the paper firmly on the photoreceptor and removing it. When the paper is pulled away, the adhesive forces strip the toner from the photoreceptor, transferring the image to the paper. The *advantage* and *disadvantage* of this method is as follows:
  - It is very efficient because it transfers toner more uniformly than electrostatic methods. Adhesive transfer is especially useful in situations where a faithful, continuous tone reproduction is needed.
  - The main problem with this method occurs in transferring extremely dense uniform images.

It should be stressed that the charge on the back side of the paper causes two effects:

1. It creates a force of attraction of the paper to the photoconductor surface and brings the paper into intimate contact with the toner particles.
2. It creates a force pulling the toner toward the paper.

The physics governing transfer is related to the competition between the force on the toner due to the electric field caused by the charge on the paper and the adhesion force of the toner to the photoconductor. The adhesion seems to be due to image forces.

4.6.5 Fusing or Fixing

The toner image is permanently fixed to the paper, preferably by heat and pressure.

The toner particles used in the development of the electrostatic image are usually made from resins, which have low melting points. These resins blend with colored pigments to get their color. The toner image is fixed to the paper by heating the paper until the toner begins to soften and flow. The toner particles first wet and coalesce, then wet and fuse to the paper. The toner migrates a small distance into the paper and forms a permanent bond.

There are also other, less common, methods of fixing:

- The toner can be fused by solvents and solvent vapors. In this case, the vapors chemically soften the resin toner particles until they bond together and to the paper. As the solvents evaporate, the toner hardens again, forming the permanent image.
- The image can be coated with a transparent lacquer, or the toner can be forced into the paper at high pressures.
- One novel method employs toner particles with encapsulated ink. The encapsulated ink is released by crushing the toner particles at high pressures between rollers. The released ink stains the paper in the form of the image.

The various methods of fusing usually require the use of heat to soften the toner and allow it to melt and flow into the paper fibers. The toner's thermal properties are a compromise between being soft for the fuser and being hard enough that they are not fused in the development system, on the photoreceptor, or in storage in a hot environment. Roll fusing (squashing the toner on paper between two rollers at least one of which is hot) requires polymeric materials to which the toner will not stick. These materials must be able to withstand high temperatures (180°C) and stress for many hundreds of thousands of copies, a nontrivial requirement for polymers. The rheological properties of the toner must be optimized to allow coalescence, spreading, and finally penetration into the paper. After fusing, the black toner decreases the reflectance of the paper, giving the perceived blackness of the image.

4.6.6 Cleaning

The photoconductor is discharged and cleaned of any excess toner using coronas, lamps, brushes, and scraper blades.

All toner that has not been transferred and all excess charge must be removed from the photoreceptor before the process starts again. Several subsystems are generally used:

- Light to discharge the photoconductor.
- AC coronas to bring the toner charge near zero.
- A brush to wipe off the excess toner.

In addition, a system to clean the cleaner is also necessary. The physics of the cleaning process involve the tension between toner adhesion to the photoreceptor and the forces on the toner caused by the cleaner. The simplest method of cleaning the photoreceptor of toner that has not been transferred is to scrape the photoconductor with a blade. This method is nearly ideal. It is simple, inexpensive, and very compact. If usage is high, the blade and the photoconductor must be changed more frequently.

The most common cleaning method is to wipe the surface with a fur brush or oscillating blade. Once the brush (or blade) frees the toner from the surface, the free toner is removed from the system by vacuum suction. It is also possible to clean the photoreceptor by cascading a granular cleaner over the residual image. The granular cleaner attracts the residual toner in a manner similar to the triboelectric attraction of the toner to the carrier beads.

4.6.7 Erasure

The erasure of the latent electrostatic image is the final step in the xerographic process. This is achieved simply by uniformly exposing the photoreceptor to an erasure lamp. The only requirement for this step is that the erasure exposure should be large enough to drive the potential to zero over the entire surface. It is important not to overexpose the photoreceptor because this, in turn, increases the internal erasure exposure of most xerographic photoreceptors. As a result, this can cause reproducibility problems if the photoreceptor is exploited (cycled) frequently.

These are the seven sequential steps that form the xerographic process. Each of these steps affects the character of the output image. We believe that two of these steps dominate system performance and are the heart of the xerographic process. These are the *latent image formation* and *development* steps.

This has been a short explanation of the main steps of the xerographic process and the different types of development systems. Although the relevant underlying physics has been identified, a complete quantitative explanation is not possible since our knowledge of the physical processes of amorphous semiconductors is not yet complete.

4.7 Photoreceptor Material Requirements

In the early stages of xerography, the only criterion used in the search for new photoconductive materials was *high dark resistance*. As it was assumed, the resistance had to be large enough so that the dielectric relaxation time exceeded the development time. Somewhat later, it became clear that the dielectric relaxation time requirement was neither necessary nor sufficient. Other factors were seen as more decisive. Instead of high-resistance conditions, a high carrier range, $\mu\tau$, and low thermal generation, G_{th}, were more important factors. In addition, relatively high carrier mobility and low equilibrium carrier concentration, $G_{\text{th}}\tau$, were also required. Since the electrical conductivity is a measure of the product (*free carrier density* $\times$ *mobility*), it provides no assurance by itself that the individual factors will fall in the correct range. In addition, the photoconductor must have blocking contacts and all of the thermal carriers must be extracted from the thickness of the material. In another formulation, xerographic photoreceptors must be charge-depletion devices with a depletion layer thickness greater than those of the device.

It should be emphasized that a good (if not a perfect) xerographic photoreceptor must:

- *be free of bulk traps,*
- *have a high mobility in order to permit the formation of the latent image in the available process time.*

Naturally, the real materials are not ideal in this context. Nevertheless, we can undoubtedly use them as photoreceptors due to their relatively satisfactory physical properties. In order to discuss requirements that the photoreceptor must satisfy in order to preserve the latent image long enough for development, we will discuss the additional dark electrical requirements.

Dark requirements:

- The free surface of the photoreceptor must contain surface traps of sufficient depth and density to hold the corona-deposited charge. These traps must be deep enough so that their mean release time is longer than the process of development time.
- The substrate must form a blocking contact for injection of the carrier, which should be opposite in polarity to the corona charge. This is necessary only for bipolar photoreceptors, such as amorphous selenium. In the case of amorphous As_2Se_3, carrier of the opposite polarity is not mobile, and the substrate contact is less important. Even with a blocking contact, the injection current I_i must satisfy $I_i\tau_{dev} \ll \epsilon E_0$, where τ_{dev} is the development time.
- The total number of free carriers thermally generated throughout the bulk of the photoreceptor must be small compared to the number on the surface $EG_{th}\tau_{dev} \ll \epsilon_s E_0$, where $\epsilon_s E_0$ is the surface charge density, G_{th} is the bulk thermal generation rate, and τ_{dev} is the time from sensitization to completion of the development.

Regarding the charging requirements, we note that the full photoreceptor thickness cannot be effective in coupling the latent image fields to development unless $Gn_r s < \sigma_s = \epsilon_s E_0$. In this expression, n_r is the bulk-free carrier concentration for the dominant charge carrier. If Gn_r is larger than σ_s, and if σ_s is as large as the material can sustain without electrical breakdown, the photoreceptor will charge with a depletion layer in the dark. The depth of the depletion layer can easily be estimated and includes nonuniform fields associated with the non-capacitive charging of the material.

The reader can find extensive examination of photo requirements, transfer versus trapping, and dark requirements in Refs. [2,3,6–8]. We may then consider the initial dark discharge current density J_d (as a quantitative measure of thermally generated discharge process). It is necessary to be specific about the nature of the physical processes contributing to J_d. In terms of the traditional dielectric relaxation time, it must be assumed that the contacts are ohmic. In other words, ohmic contacts and bulk-limited dark discharge of a photoreceptor necessarily mean complete discharge in one transit time or less. Moreover, with charging by corona devices, we conclude that a blocking contact is a necessary condition for efficient latent image formation in xerography. In practice, contacts are never perfectly blocking. Therefore, we come again to J_d replaced by an injection current density per unit area [7].

In fact, the first clue is that the required thickness of the photoreceptor must be determined by the electric field required to drive development. As assumed by Schmidlin [7], if slow trap concentration were extended into the bulk of the photoconductor, it would result in a range limitation on the discharge process, which is $\mu\tau E < L$.

Hole injection from a corona-charged surface of amorphous selenium is comparable to the bulk generation rate for a 60 μm thick sample. It is important to note here that without prior surface treatment (with nigrosine) or aging (e.g., by oxidation), the hole injection rate from the corona-charged surface of a-Se is much larger. Without any surface treatment, the total dark decay may cause the amorphous Se, as a photoreceptor, to be inoperative.

Yet the situation is not as pessimistic as it seems. It is possible to rule out electron injection from the substrate—negatively charged samples fail to discharge with strongly absorbed light. It is a well-known fact that the discharge properties of amorphous Se are affected by impurities. Electron injection from the substrate takes place since charge acceptance is usually sensitive to the material of the substrate.

The specific effect of the substrate on the total discharge rate of a-Se is known. Oxide layers thicker than 15–20 Å on aluminum provide adequate substrates for both polarities [3]. However, they work in a positively charged mode. It seems reasonable that one would use the oxide as a very large barrier to prevent the escape of holes and electrons. This could lead from cycling to enhanced electron injection. Charge fatigue (a slightly enhanced dark decay on cycling) is observed for oxide-coated aluminum substrates, and charge acceptance is limited to values corresponding to an electric field of $2 \times 10^7 \, \mathrm{V\,m^{-1}}$. This is understandable if one supposes the probability of capture of an electron by a hole, which is accumulated inside the oxide, to be below unity at the maximum charging field.

We can learn more and more about the distribution of deep traps and the surface injection properties of a-Se and a-As_2Se_3 from the study of dark discharge rate—its time, temperature, field, and thickness dependency.

4.8 References

1. C.F. Carlson, Electrophotography, US Patent 2, 297, 691, (1942).
2. R.M. Schaffert, Electrophotography Revised ed. (Focal Press, London, 1965), (Halstead Press, New York, NY, 1975).
3. J.H. Dessauer, H.E. Clark, (eds.) Xerography and Related Processes (Focal Press, London, 1965).
4. R.B. Commizzoli, G.S. Lozler, D.A. Ross, Electrophotography a Review, Proc. IEEE, 60 (1972) 348.

5. Special Issue on Electrophotographic Processes, IEEE Trans. Electron Devices, ED-19 (1972).
6. M.E. Scharfe, F.W. Schmidlin, Charged Pigment Xerography, Adv. Electron. El. Phys. 38 (1975) 83.
7. F.W. Schmidlin, in Photoconductivity and Related Phenomena (eds., G. Mort, D. Pai) (Elsevier, Amsterdam, 1976).
8. L.B. Schein, Electrophotography and Development Physics (Springer-Verlag, Berlin, New York, NY, 1992).
9. M.D. Tabak, S.W. Ing, M.E. Scharfe, IEEE Trans. Electron Devices, ED-20 (1973) 132.
10. P. Selenyi, US Patent 1, 818, 760 (1937).
11. C.F. Carlson, US Patent 2, 588, 699 (1952).
12. L. Walkup, E. Wise, US Patent 2, 618, 551 (1952).
13. R.M. Schaffert, US Patent 2, 576, 047 (1951).
14. W.E. Bixby, P.G. Andrus, L.E. Walkup, Photogr. Eng. 5 (1954) 195.
15. R.M. Schaffert, Electrophotography (Focal Press, London, 1965).
16. S.M. Vaezi-Nejad, C. Juhasz, Int. J. Electron. 67 (1989) 437.

5 Xerographic Spectroscopy of Gap States: Se-Rich Amorphous Semiconductors

Xerographic techniques, initially developed to characterize photoreceptors, are widely applicable to the study of amorphous and photoconductive insulators in relatively thin and thick film forms. Initially, the high electric field (10^5–$10^6\,V\,cm^{-1}$) due to corona charging is applied to sample films, and then the decay of the open-circuit surface potential is measured. The xerographic probe technique is a unique means to characterize electronic gap states. In particular, a map of states near mid-gap is determined by time-resolved analysis of the xerographic surface potential.

5.1 Xerographic Technique for Deep State Spectroscopy

The TOF methods are particularly important, but they cannot give sufficient information about the whole spectrum of states in the mobility gap of amorphous chalcogenides. Therefore, in addition to conventional TOF, other complimentary techniques which probe the gap states are needed, including delayed field time-of-flight (DFTOF), xerographic time-of-flight (XTOF), IFTOF [1–5], thermally stimulated conductivity (TSC), and thermally stimulated depolarization currents (TSDC) [6–16]. Xerographic techniques, which were initially developed to characterize properties of electrophotographic (xerographic) photoreceptors [17], seemed to be direct, informative, powerful, and widely applicable for the study of amorphous thin films and photoconductive insulator thin films [18].

To investigate charge transport in photoreceptor materials, conventional TOF and its various modifications have been used. Next, we focus our attention on the xerographic discharge technique. In the initial stage of this experimental technique, it is necessary to charge the surface of the examined material to a relatively high voltage. For this purpose, corona discharge is used. Therefore, before considering the principle of xerographic spectroscopy technique, we describe the construction and performance of different types of corona devices.

Amorphous Chalcogenides. DOI: 10.1016/B978-0-12-388429-9.00005-4

5.2 Corona Devices

When a high voltage of several kilovolts is applied to a corona emitter, the field near the emitter exceeds the threshold field for air breakdown. Under this condition, the molecules near the emitter become ionized. The ions may carry positive or negative charges. This depends on whether the emitter is at a positive or negative voltage. If insulating material, which is supported by earthed metal, is placed close to the corona emitter, some of the generated ions migrate to the floating surface of the sample and increase or decrease the surface charge density.

To improve the charging performance, the grounded electrodes are placed around the corona emitter. The structure called the corona housing helps direct the ion current and holds the corona emitter. A screen grid called the corona grid is inserted between the corona emitter and the sample to provide better control of charging. Several variables are known which determine the efficiency of sample charging. Among these, the most important are as follows:

- The uniformity of the charge deposited on the sample surface.
- The amount of corona current which reaches the sample surface.

The assessment of electrostatic characteristics of the sample examined depends on its particular applications. For electrophotography, the following characteristics are important: measurement of the charge acceptance (or the maximum surface charging potential), dark decay of the surface potential of the sample charged in darkness, light decay or photoinduced discharge (PID), and so-called residual potential (in other words, the potential which remains on the sample surface after discharge).

There are two types of corona devices—a pin corona discharge device and a wire corona discharge device. These devices are well described by Vaezi-Nejad [19].

A pin corona discharge device, shown in Figure 5.1, consists of a stainless steel rod with a sharp conical tip as the corona emitter, a brass hollow cylinder as the corona case, and a Perspex cup as the corona emitter holder.

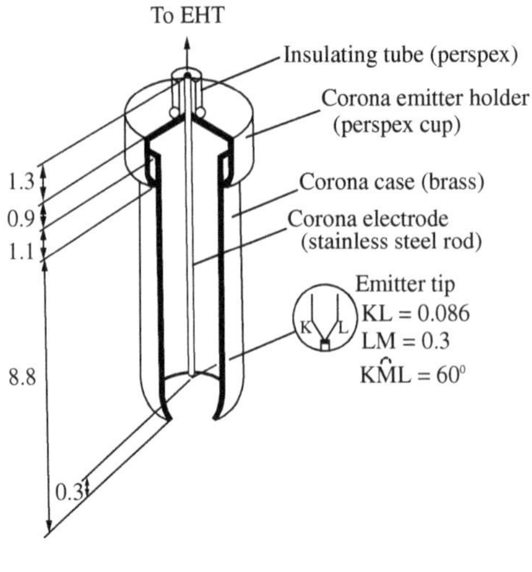

Figure 5.1 Schematic presentation of pin corona (dimensions in centimeters) [19].

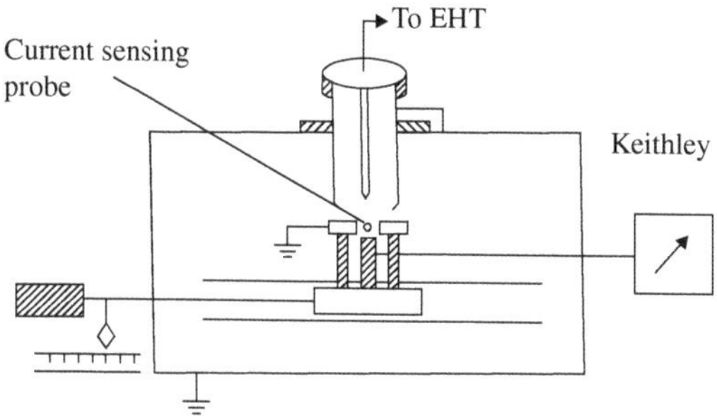

Figure 5.2 An apparatus for current distribution measurements.

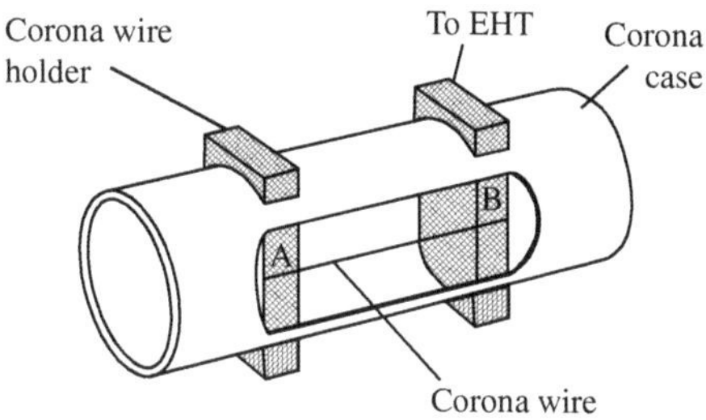

Figure 5.3 Schematic presentation of a single wire corona. Characteristics and parameters: cylindrical case diameter = 3.5 cm, length of wire = 3.7 cm, wire diameter = 76 μm, wire to window spacing = 1.0 cm, wire to back of the case distance = 2.5 cm.

Corona devices of this type were used in the past [17], but it later became clear that the electrostatic charging performance of the pin corona was unsatisfactory.

To measure the corona current distribution, the experimental apparatus of the type shown in Figure 5.2 was devised.

The most important component of this apparatus is the current sensing probe. This is a small aluminum disc (area = 1.77 mm^2) placed in the center of a relatively large plate (area = 4.2 × 2.6 cm^2). The detecting disc is electrically insulated from the plate by a 0.75 mm air gap. As a disadvantage of a pin corona device, it should be mentioned that the current distribution is not uniform.

An alternative to the pin corona is the single wire corona (Figure 5.3).

A wire corona basically consists of a hollow stainless steel cylinder (inner diameter = 3–5 cm, length = 10.7 cm, and thickness = 0.16 cm) as the corona case with a 5.7 × 3.0 cm window. Two Perspex slabs of dimensions 5.5 × 3.4 × 1.25 cm were used to hold the corona wire. The primary requirement for a corona wire emitter is that it should produce ions by means of a corona discharge at a reasonable voltage, which is typically 3–7 kV. In order to achieve high electric field, the wire diameter must be small (less than 100 μm). As the material for the corona emitter, tungsten is particularly suitable because of its resistance to the harsh environment created by the corona discharge. UV light and various nitrogen–oxygen compounds are produced in the discharge. Another advantage of tungsten is that it is mechanically strong enough to resist breakage caused by the stress of stringing, handling, and cleaning. The corona onset voltage of 1 kV is easily achievable. It is possible to improve a single wire corona's charging characteristics by adding an identical tungsten wire to the device, thus forming a double wire corona discharge. The wires are 3.7 cm long and spaced by 0.5 cm from one another [19].

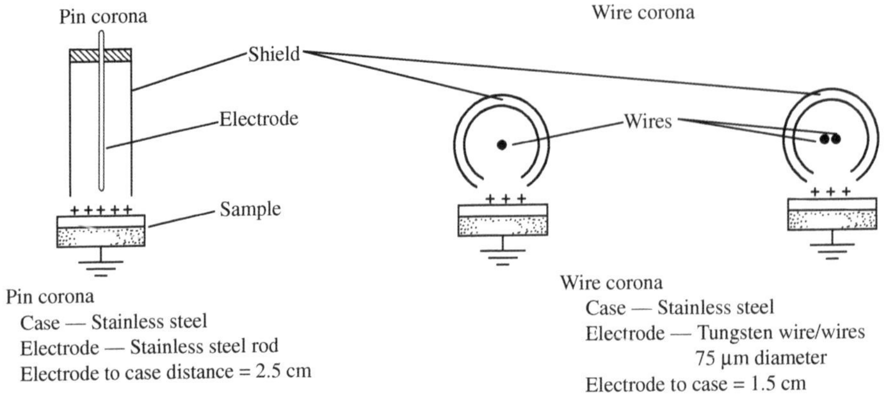

Figure 5.4 Corona devices with different configurations [20].

To find the most efficient corona discharge device for xerographic spectroscopic purposes, it is necessary to examine various device configurations utilizing sharp pins and wires as the corona emitter, as Vaezi-Nejad did [20]. The schematic sketches of different corona devices are shown in Figure 5.4.

Experimental results obtained [20] clearly illustrate that devices based on thin wires provide a more uniform charge distribution and thus may be recommended for XTOF spectroscopy. In order to control the sample initial voltage, a biased grid is inserted between the corona emitter and the floating surface of the sample.

After charging the sample, it is necessary to exactly measure the surface potential at various stages, as follows:

- *The initial voltage* of the sample for calculation of drift mobility.
- *The sample voltage after photoexcitation.* One can estimate the total charge injected into the sample. This can be made subtracting the sample voltage (after photoexcitation) from the initial voltage.
- *The final value of the sample voltage*, which is measured after photoexcitation. This is known in the literature as the residual voltage, and it is due to photoinjected carriers that are deeply trapped when they transit across the sample. If the sample is repeatedly charged and discharged, *the residual voltage* builds up with the number of cycles and saturates. In this case, we deal with *the saturated residual voltage* and can extract information about the deep traps.

After charging, the decay of the open-circuit surface potential is measured. From these measurements, important information can be extracted. Over the past decades, the xerographic probe technique became a very popular and unique means of characterizing electronic gap states. In particular, a map of states near mid-gap is determined by time-resolved analysis of the xerographic surface potential [21,22].

5.3 Principle of Xerographic Measurements Technique

An optimal photoreceptor design will require, among many other factors, high-charge acceptance, slow dark discharge, low first and cycle-up (saturated) residual voltages, and long carrier ranges ($\mu\tau$).

In general, the dynamic behavior of electrophotographic potentials is developed at two important stages in the xerographic cycle: immediately after charging in the dark (dark decay) and then immediately after photodischarge (i.e., residual decay). Both stages are extraordinarily sensitive to bulk space-charge fluctuations. In a typical sample film, $10^{12}\,cm^{-3}$ uniformly trapped electrons can give rise to several volts of surface potential, which is easily measured. From xerographic measurements, especially time-resolved analyses, information can be extracted about electronic gap states that determined the photoelectronic behavior of amorphous chalcogenides.

There are essentially three important types of xerographic behavior. They are generally termed as follows:

1. dark discharge,
2. first-cycle residual,
3. cycle-up residual voltage.

All of these must be considered when evaluating the electrophotographic properties of a-Se and its alloys. It should be stressed that these three parameters are extremely informative for the mapping of band-gap states.

In the sensitization-exposure part of the xerographic cycle, as illustrated in Figure 5.5, a photoreceptor film mounted on a grounded substrate is charged to a voltage V_0 by the corona, then completely photodischarged by exposure to strongly absorbed light (absorption depth $\delta \ll L$). Both the top corona contact and the substrate contact should be blocking. The resulting capacitor-like structure contains a uniform field after charging in the dark. Illumination with strongly absorbed light creates a thin sheet of electron–hole pairs, which separate under action of the field. Depending on charging polarity, one sign of the carrier acts to neutralize the charge on the top surface while the opposite sign of the carrier drifts through the bulk toward the grounded surface. As a result of this carrier displacement, the surface voltage decays in time. This is called PID. Ideally, the surface voltage would decay to zero. What really happens is that some fraction of carriers becomes deeply trapped during transit through the bulk. Reasonably, at the end of the illumination, there is a measurable surface potential termed the residual potential V_r. Space-charge neutrality is finally re-established by the thermally stimulated process but on a longer time scale.

It should be noted that each stage of the xerographic cycle yields important information, such as the following:

- As time resolves the charge delivered to an amorphous film from a corona, one can discern injection phenomena and measure dielectric parameters.
- By measuring the temperature-dependent dark decay of surface voltage, one can determine the energy distribution of thermal generation centers. This technique or the measurements of the dark discharge depletion time, which is the time required for the bulk to generate some fixed quantity of charge under isothermal condition, allows estimation of the position of the Fermi energy. The details of depletion discharge are considered later.
- Analysis of PID can be used to determine the parameters that characterize photogeneration and transport processes.
- From analysis of residual buildup during repetitive cycling and isothermal decay of xerographic residuals after cycling ceases, it is possible to map the spectral distribution of bulk traps in an amorphous film.

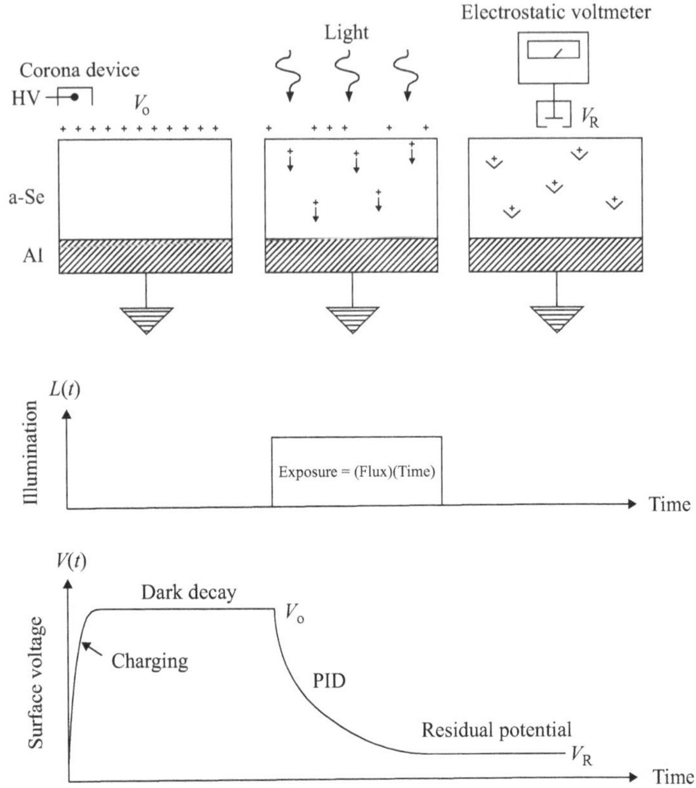

Figure 5.5 The xerographic measurements technique (schematic illustration) [18].

Corona-mode xerographic measurements were carried out using a reciprocating sample stage. In operation, the sample is first passed under a corona-charging device, the corotron, which can be set to deposit either positive or negative ions on the sample surface. The charging circuitry is such that a constant current can be supplied to the sample, simplifying the residual voltage cycle-up and in some cases, the XTOF data. After charging, the sample is moved to a measuring station, and the surface voltage is determined using a transparent, capacitively coupled electrometer probe. The time interval between termination of the charging and the onset of surface voltage measurement is a few tenths of a second. Either the dark decay, the PID, or flash illumination with strongly absorbed light can be measured. After discharge, decay of residual potential can be time resolved. The entire sequence of charging, photodischarge, and the residual can be continuously recycled. As a consequence, one can observe the progressive, stepwise buildup of residual voltage. The contacts are blocking in the sense that they allow the samples to charge capacitively in the dark for both positive and negative corona.

Dark discharge rate must be sufficiently low to maintain ample amounts of charge on the photoreceptor during the exposure and development steps. A high dark decay rate will limit the available contrast potential. The residual potential remaining after the xerographic cycle must be small enough not to impair the quality of the electrostatic image in the next cycle. Over many cycles, the cycle-up residual potential should be small to avoid deterioration in the copy quality after many cycles.

In the case of a-Se, these xerographic properties have been extensively studied by numerous authors. In addition to the magnitude of the saturated residual voltage, the rate of decay and the temperature dependence of the cycle-up residual potential are important considerations, since they determine the time required for the photoreceptor to retain its first-cycle xerographic properties.

Figure 5.6 displays the simplest experimental setup for xerographic measurements. The rotating photoreceptor drum is charged at station A by a corotron device. The surface potential is measured at B, and the photoreceptor is then exposed to a controlled wavelength and intensity illumination at station C, following which its surface potential is measured again at station D. In some systems, the surface potential is also monitored during exposure at C via a transparent electrometer probe to study the PID characteristics. Normally, the charging voltage, speed of rotation, and exposure parameters, such as energy and wavelength, are user-adjustable (Figure 5.7).

The residual potential can be related to the space-charge density. For the case of uniform bulk space-charge of density ρ_0, we may write [21] the simplified expression for residual potential V_r as

$$V_r = \rho_0 L^2/2\varepsilon_0\varepsilon = NeL^2/2\varepsilon_0\varepsilon, \tag{5.1}$$

where L is the sample thickness, ε is the relative dielectric constant, ε_0 the free-space permittivity, N is the number per unit volume of electronic charges e. A relatively

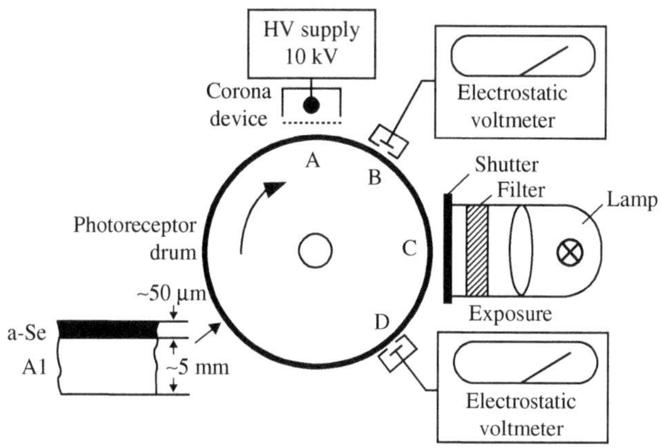

Figure 5.6 Simplified schematic diagram of a xerographic measurement. The photoreceptor is charged at A and exposed at C. Its surface potential is measured before and after exposure at B and D [18].

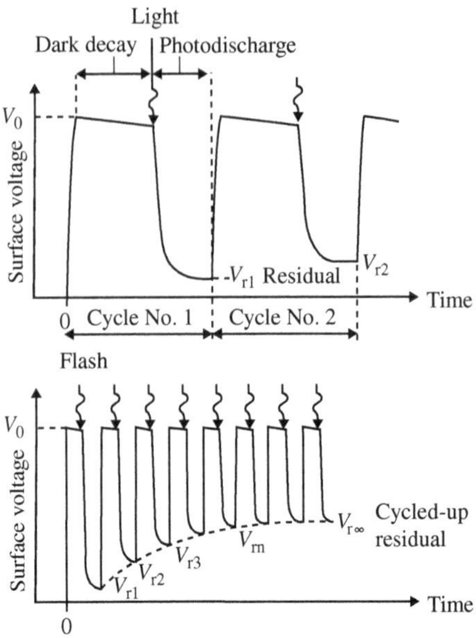

Figure 5.7 Typical photoreceptor behavior through xerographic cycles showing dark decay, first-cycle residual potential V_{r1}, and cycle-up residual potential V_{rn} after many cycles [18].

small density of surface or bulk-trapped charges on or in typical xerographic film can give rise to an appreciable surface voltage (Figure 5.8) [21].

The $\mu\tau$ product, where μ is the drift mobility and τ the bulk deep trapping lifetime, can be determined from the first-cycle residual potential of a well-rested sample of thickness L. In the weak trapping limit (when the residual V_r is much less than the charging voltage V_0), Kanazawa and Batra [23] derived the expression

$$V_r/V_0 = (0.5L^2/\mu\tau V_0)[-\ln(2V_r/V_0)]. \tag{5.2}$$

The physically plausible expression derived by taking the residual to be that voltage for which the carrier range is nominally half the sample thickness is

$$V_r/V_0 \approx 0.5L^2/\mu\tau V_0. \tag{5.3}$$

In the strong trapping limit (V_r/V_0), Kanazawa and Batra [23] derived

$$V_r/V_0 = 1 - (\mu\tau V_0/L^2)(\ln 2), \tag{5.4}$$

which differs only slightly from the expression based on the previous physical definition of residual

$$V_r/V_0 = 1 - (\mu\tau V_0/L^2). \tag{5.5}$$

It should be noted that bulk deep trapping lifetimes computed from first-cycle residuals are in agreement with lifetimes measured in the TOF mode under

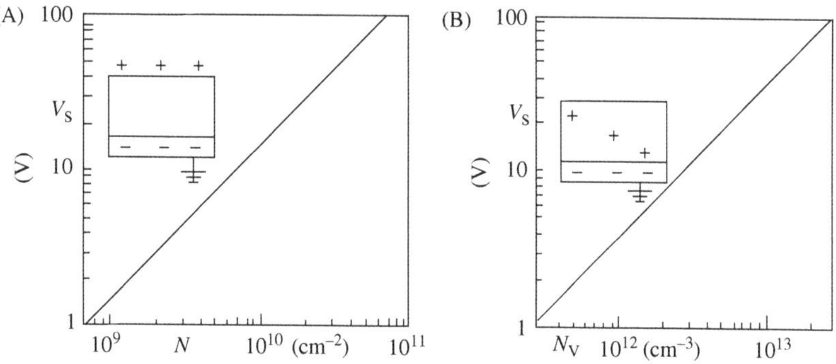

Figure 5.8(A) Surface voltage V_s versus surface density of electronic charges. (B) Surface voltage V_s versus volume density of electronic charges (bottom). Both computed using Eq. (5.1) for a 50-μm- thick film, $\varepsilon = 6.4$ [21].

range-limited conditions. For example [21], in a-Se specimen with $L = 48\,\mu$m charged to 200V, hole and electron residuals at $T = 295$ K were found to be 1.8 and 44V, respectively. The measured drift mobility, lifetimes, and $\mu\tau$ products, for electrons are, respectively: $4.9 \times 10^{-3}\,\text{cm}^2\text{V}^{-1}\text{s}^{-1}$, 5×10^{-5} s, and $2.45 \times 10^{-7}\,\text{cm}^2\text{V}^{-1}$. The measured drift mobility, lifetimes, and $\mu\tau$ products for holes are, respectively: $0.16\,\text{cm}^2\text{V}^{-1}\text{s}^{-1}$, 4.4×10^{-5} s, and $7.04 \times 10^{-6}\,\text{cm}^2\text{V}^{-1}$. The $\mu\tau$ product computed using Eq. (5.4) is $2.62 \times 10^{-7}\,\text{cm}^2\text{V}^{-1}$ for electrons and $6.4 \times 10^{-6}\,\text{cm}^2\text{V}^{-1}$ for holes [21].

5.4 Dark Discharge in a-Se

The surface potential of a charged photoreceptor will decay even in the dark. Figure 5.9 shows typical positive and negative dark discharge curves for pure a-Se films prepared under different conditions. The dark discharge rate depends on the substrate temperature [17].

In principle, thermal generation of carriers in the bulk, hole, and electron transport and interfacial injection can all be the cause of dark decay. The simplest picture of dark decay is one in which electron–hole pairs are thermally generated in the bulk at a rate G_B and swept rapidly compared with the generation rate [24]. The time-dependent field E can be obtained by setting the sum of the displacement and conduction currents equal to zero

$$\frac{dE}{dt} = e(J_s + G_B L)/\varepsilon, \tag{5.6}$$

where J_s is a surface charge generation rate ($\text{cm}^{-2}\text{s}^{-1}$), ε is the bulk dielectric constant, and L is the film thickness. As indicated in Eq. (5.7), a series of measurements of samples of varying thickness are required to distinguish bulk from surface contribution to dark decay [25].

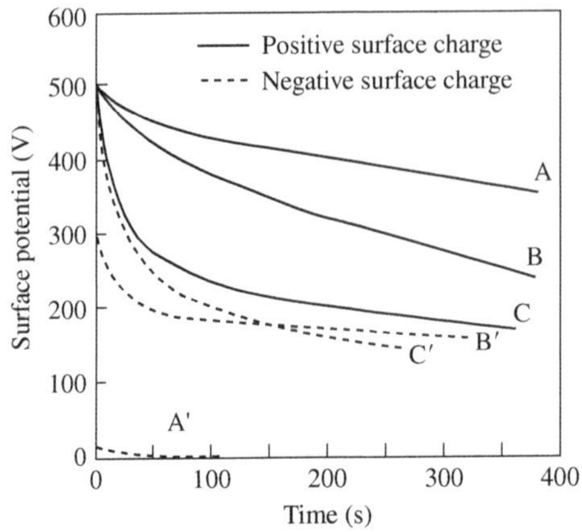

Figure 5.9 Dark decay of surface potential on a-Se layers. A, B, and C involve a-Se deposited under different substrate temperature (T_s) conditions: A and A′ at $T_s = 75°C$; B and B′ at $T_s = 50–60°C$; C and C′ at $T_s = 25–50°C$ and uncontrolled [18].

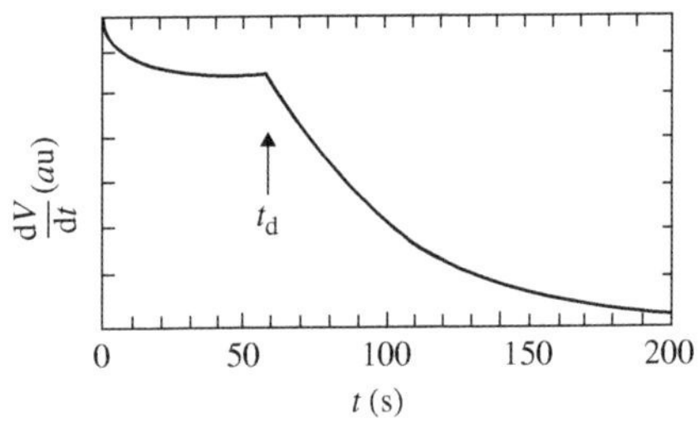

Figure 5.10 Typical dark decay (d*V*/d*t* versus *t*) for a-Se film doped with 750 ppmw Cl measured immediately after charging in the dark to 75 V. Thickness is 15 μm.

The dark decay in pure and alloyed a-Se films may be caused by all of several factors: substrate injection, bulk thermal generation, and depletion. It has been found that the depletion-discharge mechanism dominates. At present, this mechanism is generally accepted [26–32].

Experimental studies of dark discharge performed on various Se-based alloys clearly illustrate that instead of the type of behavior predicted by Eq. (5.6), dark decay is characterized by two distinct zones of time dependence separated by an abrupt transition (Figure 5.10).

This feature can be completely accounted for by a model which presumes that only one sign of carrier is mobile. During dark decay, the mobile carrier is depleted (in other words swept out), leaving behind a space-charge of opposite sign. Therefore, the dark decay is called a depletion discharge.

In the following, we consider the depletion-discharge model, which describes a dark discharge where only one sign of thermally generated charge carrier is mobile on the time scale of the experiment. As thermal generation and sweep-out of the mobile carrier proceeds, deeply trapped space-charge $\rho(z)$ of the opposite sign is developed. The time-dependent surface voltage is computed at any time by integrating the instantaneous field across the sample thickness, taking into account all of the sources. This procedure directly allows one to distinguish the bulk charge generation process from surface charge losses.

The instantaneous surface potential is given by [25]

$$V = \int_0^{Z_0} \mathrm{d}z E(z) = \frac{1}{\varepsilon}\int_0^{Z_0}\left[\mathrm{d}z\left(\sigma - \int_0^{Z_0} \mathrm{d}z'\rho(z)\right)\right], \tag{5.7}$$

where σ is the surface charge density initially deposited minus any loss by injection up to the instant of measurement. Note that Z_0 is the depth (measured from the top surface) at which $E(Z_0)$ is zero and is equal to or less than thickness L. At $Z_0 < L$, the surface charge density and Z_0 are related by

$$\sigma = \int_0^{Z_0} \rho(z)\mathrm{d}z. \tag{5.8}$$

If depletion is a spatially homogeneous process, Eqs (5.7) and (5.8) simplify and the instantaneous surface voltage is then given by

$$V = (\sigma Z_0/\varepsilon) - (\rho Z_0^2/2\varepsilon), \tag{5.9}$$

where Z_0 and ρ are time dependent.

The process seems to proceed as follows: initially, when the bulk charge is less than the surface charge ($\sigma > \rho L$), Z_0 equals L and is time independent. Later, when Z_0 is less than L, Z_0 equals σ/ρ and is therefore time dependent. As a result, with increasing ρ, the parameter Z_0 decreases. Between these zones is a demarcation defined by the condition

$$\rho L = \sigma, \tag{5.10}$$

which occurs when $t = t_\mathrm{d}$, where t_d is the depletion time. The explanation of the physical process at depletion discharge is, at first sight, simple. As the charge is uniformly depleted from the bulk of the sample, ρ increases in time until the depleted charge equals the surface charge Eq. (5.10). Further depletion now begins to reduce the depleted volume by reducing the dimension $Z_0 = \sigma/\rho < L$. As Z_0 shifts toward $Z = 0$, the region between $Z = Z_0$ and $Z = L$ retains its space-charge neutrality. The dark discharge rate $\mathrm{d}V/\mathrm{d}t$ and the surface voltage each exhibit different behavior in the two respective time zones separated by t_d. In zone 1, when $t < t_\mathrm{d}$, one obtains (at a condition where there is no surface charge loss)

$$V = \frac{\sigma L}{\varepsilon} - \frac{\rho L^2}{2\varepsilon} \tag{5.11}$$

and

$$\frac{\mathrm{d}V}{\mathrm{d}t} = -\frac{L^2}{2\varepsilon}\frac{\mathrm{d}\rho}{\mathrm{d}t}. \tag{5.12}$$

In zone 2, when $t > t_d$, we obtain

$$V = \sigma^2/2\varepsilon\rho \tag{5.13}$$

and

$$\frac{dV}{dt} = -\frac{\sigma^2}{2\varepsilon\rho^2}\frac{d\rho}{dt}. \tag{5.14}$$

From Eq. (5.11), it is clear that

$$V = V_0 = \sigma L/\varepsilon \quad (\text{at}\, t = 0)$$

and (at t_d)

$$V = \sigma L/2\varepsilon = V_0/2 \quad (\text{when}\, \rho L = \sigma). \tag{5.15}$$

Therefore, the voltage is reduced to half its initial value at the depletion time. Equations (5.12) and (5.14) are modified when surface loss by injection is included by the addition of a surface term, given by

$$\left(\frac{dV}{dt}\right)_s = \frac{L}{\varepsilon}\frac{d\sigma}{dt} \quad \text{in zone 1} \tag{5.16}$$

and

$$\left(\frac{dV}{dt}\right)_s = \frac{\sigma}{\varepsilon\rho}\frac{d\sigma}{dt} \quad \text{in zone 2}. \tag{5.17}$$

The boundary between the zones is defined by condition $\rho(t_d)L = \sigma(t_d)$. It follows that

$$V(t_d) = \frac{\sigma(t_d)L}{2\varepsilon} = \frac{\sigma(t_d)}{\sigma(0)}\frac{V(0)}{2}, \tag{5.18}$$

where $\sigma(0)$ and $V(0)$ are values at $t = 0$. It should be stressed that the extent to which the surface voltage at t_d has decayed beyond $V_0/2$ provides a quantitative measure of the surface loss in the specimen.

One representation of $\rho(t)$ is based on the assumption that emission is a first-order rate process. This applies to a system containing $N_0(\text{cm}^{-3})$ discrete emission centers displaced by energy E_0 from the transport state. For this system, $\rho(t)$ t because

$$\rho(t) = en = eN_0[1 - \exp(-Rt)], \tag{5.19}$$

where

$$R = \nu \exp(-E_0/kT) \tag{5.20}$$

and

$$\rho(t) < eN_0, \quad Rt < 1,$$

thus

$$\rho(t) \approx eN_0Rt. \tag{5.21}$$

From Eqs. (5.10), (5.20), and (5.21), it follows that for emission from a discrete center,

$$t_d = (2V_d)\varepsilon/eN_0L^2\nu \exp(-E_0/kT), \tag{5.22}$$

where we abbreviate $V_d = V(t_d)$ so that in this case a plot of $\log t_d$ versus $\log(2V_d)$ should be linear. Any curvature when the log of t_d is plotted against $\log V_0$ (V_0 = surface voltage at $t = 0$) is then a measure of surface loss by injection. Both features are clearly displayed in Figure 5.11.

Here, we note that the scaling of t_d with charging voltage as well as the shapes of the associated depletion curves are observed in a series of Se-rich amorphous alloys and also in pure selenium (in the latter case only in thick 50 μm films charged to relatively low voltages at which surface loss is negligible).

In a series of elegant experiments, Abkowitz et al. [33] demonstrate that the broadening of shallow transport interactive states identified in TOF experiments (see the increasing dispersiveness of TOF transients in Figure 5.12) occur in parallel with a similar process operating on the deep gap states which control the dark decay.

The rate at which bulk space-charge is uniformly generated may be represented as

$$\rho(t) = at^p, \quad p < 1. \tag{5.23}$$

This algebraic time dependence is suggested by the analytical modeling of simple cases. Let us consider exponential distribution of emission centers given by

$$N = N_0 \exp(-[E - E_0]/W), \tag{5.24}$$

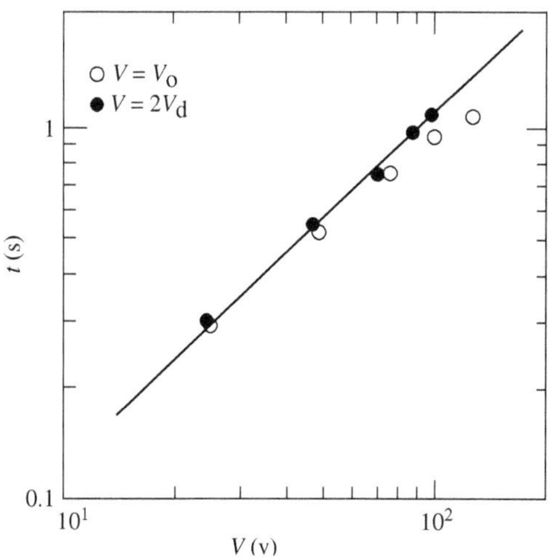

Figure 5.11 Dependence of the depletion time t_d on the charging voltage V_0 (open circles) and twice the depletion voltage $2V_d$ (solid circles) for a-Se film doped with 750 ppmw Cl. Thickness is 15 μm.

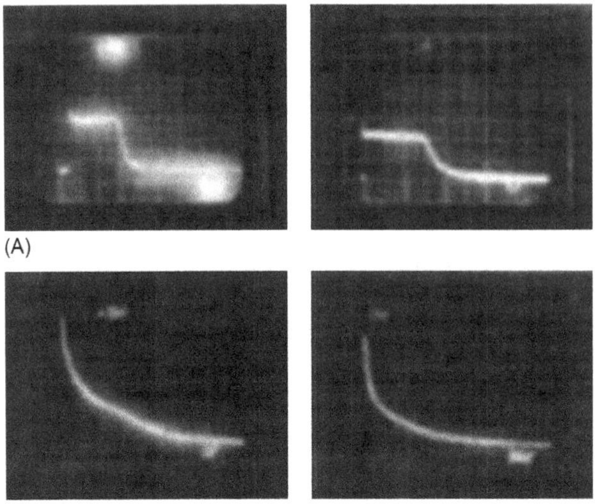

(A)

(B)

Figure 5.12 Room-temperature transit pulse shapes for holes (left-hand side) and electrons (right-hand side). (A) a-Se; $E = 10\,\mathrm{V\,\mu m^{-1}}$; $0.1\,\mathrm{\mu s\,div^{-1}}$ for holes and $2\,\mathrm{\mu s\,div^{-1}}$ for electrons. (B) 18.4 wt% Te-Se; $E = 17.5\,\mathrm{V\,\mu m^{-1}}$; $1\,\mathrm{\mu s\,div^{-1}}$ for holes and $200\,\mathrm{\mu s\,div^{-1}}$ for electrons.

which is N_0 at the reference energy E_0, and it is also cut off by the Fermi energy for holes (states extending beyond E_F are hole traps rather than emission centers).

It follows that

$$\rho(t) = eN_0(\nu t)^{kT/W} \tag{5.25}$$

Here, it is assumed that the emission rate pre-factor v does not depend on energy, and there is no re-trapping. Substituting Eq. (5.24) into corresponding equations for $\mathrm{d}V/\mathrm{d}t$ yields

$$\frac{\mathrm{d}V}{\mathrm{d}t} = -\frac{L^2}{2\varepsilon}a\rho t^{p-1}, \quad t < t_\mathrm{d} \tag{5.26}$$

and

$$\frac{\mathrm{d}V}{\mathrm{d}t} = \frac{\varepsilon}{2a}(V_0/L)^2\rho t^{-p-1}, \quad t > t_d. \tag{5.27}$$

The following distinctive characteristic are now predicted for depletion discharge:

- There is an abrupt change in the dark decay rate ($t^{2\mathrm{p}}$) at $t = t_\mathrm{d}$. The physical origin of t_d is required by the bulk to isothermally generate a quantity of charge equal to the surface charge. This parameter (t_d) provides a quantitative measure of thermal generation.
- The sum of the slopes of two linear segments corresponding to $t < t_\mathrm{d}$ and $t > t_\mathrm{d}$ equals -2.
- At $t = t_\mathrm{d}$,

$$Lat_d^p = \sigma$$

or

$$t_d = (\sigma/aL)^{1/p} = (2V_d\varepsilon/\sigma L^2)^{1/p}. \tag{5.28}$$

- When a series of films differing in thickness L are charged to the same initial field, then a transition from quadratic to thickness-independent behavior will be observed in the dark decay rate at t_d.

Features predicted by Eqs. (5.26) and (5.27) are clearly illustrated by Figure 5.13. A demarcation in the rate of decay is now easily discerned in all cases.

In the depletion-discharge model, both the shapes of the dark decay and the scaling of the depletion time t_d occur with the depletion voltage V_d (half the charging voltage in the absence of surface loss), and specimen thickness is completely specified by the p parameter. The form taken by the phenomenological depletion model resembles a key feature predicted in the model proposed by Scher and Montroll [34] for dispersive transport. In their model, it is the dispersion parameter α, which governs a similar scaling law for the transit time. In Figure 5.14, the depletion time t_d represented on the right ordinate log scale and the p parameter represented linearly on the left are plotted versus composition with samples spanning the range 0–14 at.% in Se.

It is clearly seen that under fixed conditions, the depletion time decreases exponentially with Te concentration while the p parameter decreases linearly over the same composition range.

This simple phenomenological model, which assumes a power law time-dependent development of spatially uniform negative charge density $\rho(t) = at^p$, can account for all features of the dynamic behavior of the surface potential. Thus, when plotted on a log–log scale, the dV/dt versus t curve for a wide range of alloy films was found to exhibit a well-defined kink precisely at time t_d when the negative bulk-trapped

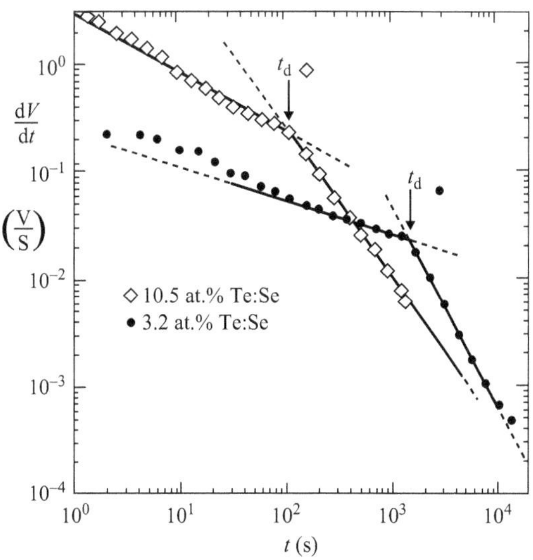

Figure 5.13 Dark decay at 297 K of a 3.2 at.% Te-Se film (solid circles) and a 10.5 at.% Te-Se film (diamonds). Films were initially charged to 100 V. Thickness is $d = 55\,\mu$m [25].

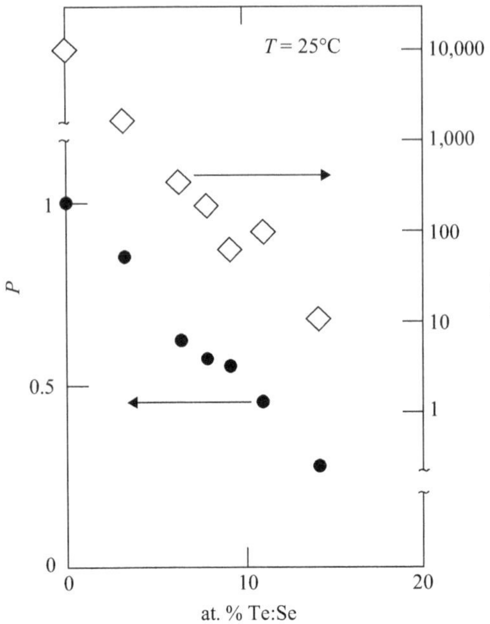

Figure 5.14 Composition dependence of the depletion time t_d and the parameter p for Te-Se alloy films [25].

space-charge became equal to the surface charge initially deposited. Depletion time t_d provides a measure of the time required by the bulk to generate an experimentally determined quantity of charge. The shape of the decay of dV/dt versus t on either side of t_d and the scaling of the depletion time t_d with surface charge σ and sample thickness L, $t_d = (\sigma/aL)^{1/p}$, respectively, provide independent means for measuring parameter p. If the buildup in the bulk negative charge density is spatially uniform (as we assume), the internal electric field falls linearly with distance from the top surface. At a certain time t_d, the electric field F at the grounded end of the sample becomes zero. From that time onward, the field will be zero at a distance $X(t) < L$, the sample thickness, and consequently there will be a neutral region from X to L, as holes generated in $0 < x < X$ and arriving into $X < x < L$ will not be swept out. The shrinkage of the depleted volume with time $t < t_d$ means that t_d marks a functional change in the dark decay rate and therefore is readily obtainable from dark discharge experiments (Figure 5.15).

Analysis performed for a range of Se-based alloys reveals that the parameter p and the depletion time t_d scale systematically with alloy composition when experimental conditions remained fixed. Parameter t_d and p decrease with an increase of the concentration of arsenic, antimony, or tellurium in the alloy [18,25,35,36]. The quantity p in the model for dark depletion discharge plays a role analogous to the dispersion parameter α in the theory of dispersive transport [34] of photoinjected carriers.

Under low-charging voltages, the depletion time indicates the time required for the surface potential to decay to half its original value. Under high-charging voltages, however, field-enhanced emissions from the deep mobility gap centers also

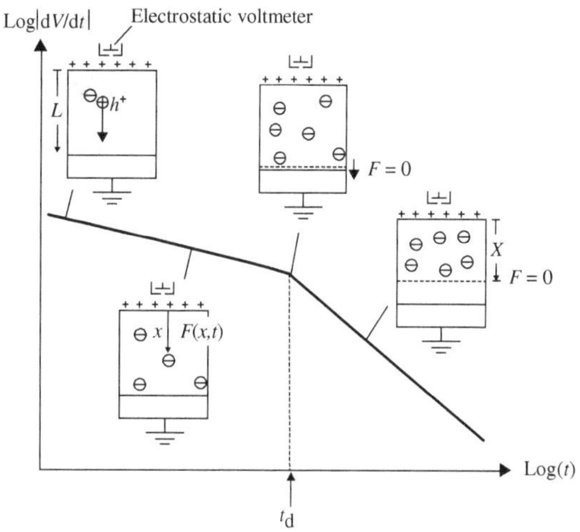

Figure 5.15 Typical log–log plot of the dark discharge rate versus time for a wide range of amorphous Se-based alloys.

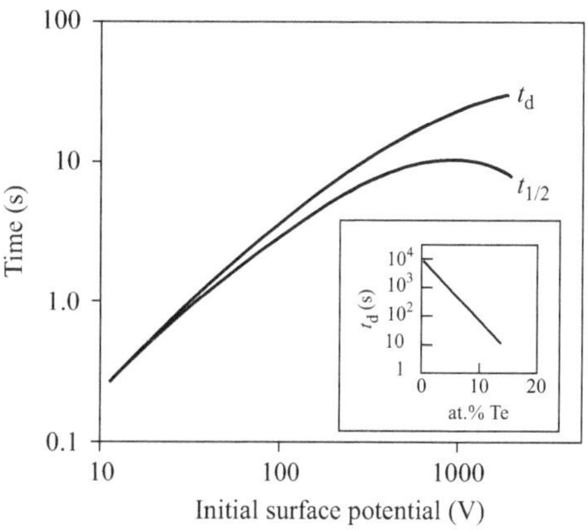

Figure 5.16 Log–log plots of the depletion time t_d and time for the surface potential to decay to its half value $t_{1/2}$ versus charging voltage V_0 for an amorphous Se–Te 13 wt% photoreceptor film of thickness 70 μm [30]. The inset shows the dependence of the depletion time t_d on the Te content [31].

play an important role. The surface potential initially decays at a much faster rate so that at the depletion time, the surface potential is in fact less than half the initial value. Figure 5.16 shows the dependence of the depletion time t_d and the half time $t_{1/2}$ on the charging voltage V_0, where it can be seen that at the highest charging voltages there is no improvement in $t_{1/2}$ with further increase in the charging voltage V_0. Inasmuch as the dark decay in Se-based alloys is a bulk process, the rate of discharge increases with the square of thickness $dV/dt\ L^2$ and can be reduced only by using thin layers. This concept leads naturally to the design of multilayer photoreceptor structures.

The origin of the deep localized states in the mobility gap, which control the dark decay, has been attributed to structural native thermodynamic defects [32,37–39]. Thermal cycling experiments show that the response of the depletion time to temperature steps is retarded, as would be expected when the structure relaxes toward its metastable, liquidlike equilibrium state. As the structure relaxes toward the equilibrium state, t_d decreases further until the structure has equilibrated. The only possible

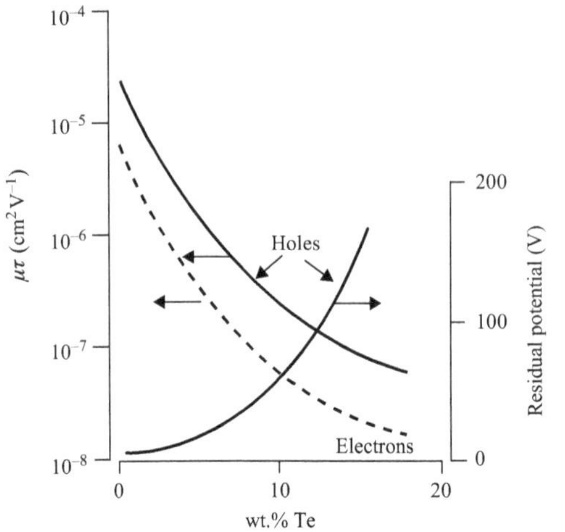

Figure 5.17 Hole and electron drift mobility lifetime product $\mu\tau$ and residual potential versus Te content in a-$Se_{1-x}Te_x$ films. The $\mu\tau$ product was xerographically measured by Abkowitz and Markovics [33]; residual potential can be seen in Ref. [40].

inference is that t_d must be controlled by structure-related thermodynamic defects. The generation of such defects is therefore thermally activated. We should note that since the depletion-discharge mechanism involves the thermal emission of carriers from deep localized states, it is strongly temperature dependent. For example, t_d increases in an approximate Arrhenian fashion with decreasing temperature. In addition to the deterioration of the dark decay, there is an increase in the residual potential for a-$Se_{1-x}Te_x$ alloys.

Figure 5.17 displays the $\mu\tau$ product for holes and electrons [33].

This parameter was determined from xerographic residual potential in a-$Se_{1-x}Te_x$ monolayer films. As one can see, even with very little Te alloying, there is a considerable rise in both hole and electron deep traps. The relationship between the trapping time and the residual potential has been evaluated by numerous authors (e.g., [23,41]). The residual potential is more than an order of magnitude larger than typical values for pure selenium in the situation when the Te concentration exceeds 12 wt% Te.

Regarding the depletion discharge it should be stressed that with thick films, a good blocking contact between a-Se and the pre-oxidized aluminum substrate, the depletion mechanism dominates the discharge process. There are several reasons for the experimentally observed fact that amorphous selenium possesses good dark decay characteristics. Among these are the following:

- There are not many deep localized states in the mobility gap of pure amorphous selenium. The concentration of such states is relatively low (less than 10^{12}–10^{13} cm^{-3}).
- The energy location of these localized states is deep (>0.8 eV) in the mobility gap. Therefore, the thermal generation process of holes and/or electrons from these centers is slow.
- Injection from the substrate can be reduced substantially by using oxidized Al substrate.

If blue light is used for the discharge process, then the absorption is very close to the charged surface, and one can assume that the discharge process involves the transport of photogenerated holes through the bulk. Trapping of these holes in the bulk then results in the observed first-cycle residual potential V_{r1}. In the case of amorphous selenium photoreceptor films, it has been found that V_{r1} is predicted by the Warter [42] expression, $V_{r1} = L^2/(2\mu_d\tau_0)$, where L is the film thickness, μ_d is the hole drift mobility, τ_0 is the hole lifetime, and $\mu_d\tau_0$ is the hole range.

It can be seen in Figure 5.7 that as the xerographic cycle is repeated many times at a constant repetition frequency, the residual voltage rises and eventually saturates. The saturated residual voltage V_{rs} is much larger than the first-cycle residual voltage V_{r1}. Both the first residual and the cycle-up saturated residual potential, V_{r1} and V_{rs}, are sensitive to pre-illumination as well as to temperature and alloying. For example, when a-Se films are pre-illuminated with white light, the buildup of the residual potential occurs more rapidly toward a much higher saturated residual potential. Furthermore, the parameters mentioned (residual potentials V_{r1} and V_{rs}) increase with exposure time. The effects of pre-illumination with band-gap light on the main xerographic characteristics in the case of pure amorphous selenium were examined in detail by Abkowitz and Enck in their extensive work [22]. For Se-rich a-As_xSe_{1-x} alloys, see Refs. [43,44] by the authors of this book. Clearly, exposure to white light generates an appreciable concentration of deep hole traps. The effects of pre-illumination are considered next.

There are two reasonable explanations for the saturation of the residual voltage [45]. These are as follows:

1. The observed saturation may be due to the dynamic balance between trapping and release of charge carriers as the xerographic cycle is repeated.
2. As an alternative, it may be due to the filling of the deep trap population so that the saturated residual potential is given by $V_{rs} = L^2eN_t/(2\varepsilon_0\varepsilon)$, where N_t is the concentration of deep traps and ε_0 and ε are, respectively, the absolute permittivity and relative permittivity of the photoreceptor material. The trap-filling interpretation can be verified by checking whether V_{rs} shows any dependence on the cycling frequency. This excludes a dynamic equilibrium condition involving trap filling and emptying (Figure 5.18).

The rate of decay and the temperature dependence of the saturated voltage can be used to obtain the concentration and energy distribution of the deep traps responsible for the residual potential. Thus, V_{rs} provides a useful means of studying the nature of deep traps in amorphous semiconductors and has been successfully used to derive the energy distribution of deep localized states in the mobility gap of both a-Se and a-Si–H [21,47].

One can assume that the saturated residual potential, at the end of a large number of cycles, decays. As thermal release proceeds, holes are emitted and swept out from the specimen, resulting in the decrease of the measured surface potential. The decay rate of the saturated potential is strongly temperature dependent due to thermal release from deep mobility gap centers, located at ~0.9 eV above E_v for holes. The discharge of the saturated potential due to electron trapping occurs much more slowly. The reason is that the energy depth of electron traps from E_c is about 1.2 eV, which is greater than that of hole traps from E_v.

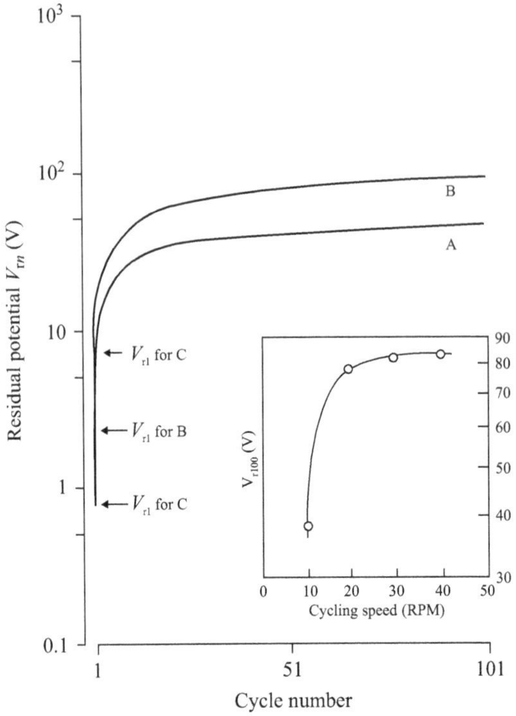

Figure 5.18 Typical cycled-up xerographic experiments on a-Se films. The inset shows the dependence of residual potential at the 100th cycle on the cycling speed. Sample A has a lower $\mu\tau$ product than B [46].

5.5 Photoinduced Changes of Xerographic Characteristics: Dark Discharge

Amorphous chalcogenide semiconductors exhibit a large spectrum of photoinduced phenomena. According to Tanaka [48], these unique phenomena observed in chalcogenide glasses can be classified into two main groups.

The first group includes the so-called heat-mode phenomena: the heat generated through nonradiative recombination of photoexcited carriers causes atomic structural changes. The most familiar may be the phase change between crystalline and amorphous phases, which is applied to high-density (more than 1 GB) erasable optical memories.

The second is the so-called photon mode. It is well known that this mode has unique effects with amorphous chalcogenide semiconductors. Reversible thermal and photoinduced changes in amorphous semiconductors have been under active investigation during several decades [48–55]. They are the most unique and intriguing features of chalcogenide vitreous semiconductors. Photoinduced changes appear as nearly parallel shifts of the optical absorption edge to lower energy or a reduction in band-gap on exposure to light (so-called photodarkening). Such irradiation also causes a change in various physical properties. The physical origin of the processes which takes place at reversible transformations induced by band-gap light and thermal annealing remains unclear. Particular interest is aroused when observations suggest that these changes affect photoelectronic behavior. It is important to note that analysis of changes in photoelectronic properties can be used to relate photoinduced effects to specific changes in the electronic density of gap states.

During the last few decades, there has been, however, very limited systematic study of photoinduced effects on states, especially of deep lying states in the band-gap. In addition, most photodarkening studies have focused upon prototypical binary (and ternary) chalcogenide alloys, and relatively little is known about the characteristics of elemental and chalcogen-rich glasses. At the same time, elemental and chalcogen-rich amorphous semiconductors serve as useful model systems for studying the influence of photodarkening on physical properties. Therefore, we consider photoinduced changes of deep levels in pure and selenium-rich amorphous layers.

Careful analysis of the time and temperature-dependent decay of the surface voltage on an amorphous film can be used to map the density of states. This procedure can be performed after corona charging of the sample but prior to exposure (known as xerographic dark decay) or after exposure (PID) in the first cycle or after many cycles. The procedure is illustrated for a-As_xSe_{1-x}, where residual potential can be measured without complication. Amorphous As_xSe_{1-x} ($0 < x < 0.20$) is found to be characterized by relatively discrete gap state structure. Our measurements readily discern photo- and thermostructural effects on gap state population [43,44,52,55]. Thus, during these structural transformations, systematic variations in the density of localized states distributed throughout the mobility gap are observed. This observation is consistent with the view that native defects play a key role in photoelectronic behavior of amorphous chalcogenides.

The illumination of amorphous As_xSe_{1-x} films by light with energy near the optical gap also causes changes in basic electrophotographic parameters: dark discharge rate, initial-charging potential, residual potential, and its dark decay rate. Typical dark discharge curves for film composition $As_{0.2}Se_{0.8}$ are shown in Figure 5.19.

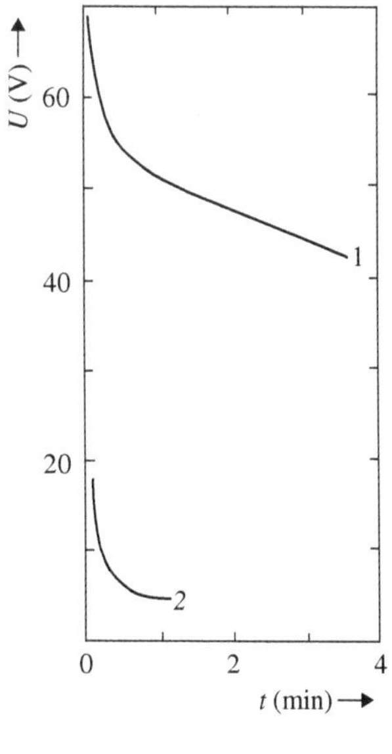

Figure 5.19 Dark decay of surface potential in a-$As_{0.2}Se_{0.8}$. (1) Dark-rested sample, (2) after white light exposure 5,000 lx × 60 s.

Table 5.1 Surface Potential Dark Decay Parameters and Their Photoinduced Changes in As_xSe_{1-x} Films

Content	dU/dt (V/s)*	$t_{U0/2}$ (s)	γ	$\Delta U/U_0$
0	6.7×10^{-2}	276	1.06	0.30
2	9.7×10^{-2}	150	1.20	–
5	1.4×10^{-1}	138	1.33	–
8	9.3×10^{-1}	19	1.61	–
10	1.22	17	2.3	–
15	1.12	18	3.5	0.50
20	1.4	15	1.9	0.51

*Initial voltage: $U_0 = 37$ V; thickness: $d = 10\,\mu$m; $T = 293$ K; white light exposure: 2,000 lx × 120 s.

It is apparent that the initial-charging voltage U_0 becomes smaller, and the surface potential decay rate dU/dt increases after previous photoexcitation. It must be emphasized that, depending on composition, ordinary (changes of either U_0 or dU/dt) or complex (simultaneous changes of U_0 and dU/dt) photoinduced effects may be observable. The former take place in the range 2–6 at.% As, whereas the latter concentrations require 8 at.% As.

In pure selenium, the photoinduced change of discharge rate is comparatively small ($\gamma \approx 1.06$). As the arsenic concentration increases, γ initially becomes progressively larger, then for $x > 20$ at.% As it begins to fall, possibly because of the rising dark discharge rate in dark-rested samples. The parameters of dark discharge in dark-rested and photosensitized films are summarized in Table 5.1. Here, $t_{U0/2}$ is the half-decay time, and $\Delta U/U_0 = (U_0 - U_0^*)/U_0$ is the relative change of the initial-charging potential. It is of interest to point out that the memory effect (the time interval during which the changes in pre-excited film parameters may be observable) is appreciably prolonged with As content: from ≈40 min in pure selenium to ≈10 h in $As_{0.2}Se_{0.8}$.

The ratio $\gamma = (\mathrm{d}U^*/\mathrm{d}t)/(\mathrm{d}U/\mathrm{d}t)$ (parameters of irradiated samples are marked by asterisk) is essentially composition dependent.

Figure 5.20 shows the time evolution of the dark decay kinetics of the surface potential after photoexcitation. An increasing dark adaptation time (i.e., the dark-resting time of an exposed film before charging it to a certain potential and recording the potential decay) causes a diminishing of the observable photoinduced changes. Compared to the photocurrent transients [43,44], the relaxation functions $\varphi(\tau) = [U_0 - U_0^{\tau}]/[U_0 - U_0^*]$ and $\zeta(\tau) = [(\mathrm{d}U^*/\mathrm{d}t) - (\mathrm{d}U^{\tau}/\mathrm{d}t)]/[(\mathrm{d}U^*/\mathrm{d}t) - (\mathrm{d}U/\mathrm{d}t)]$ were estimated (see inset in Figure 5.20). This characterizes the initial-charging potential and the dark decay rate recovery.

It is of particular significance that memory effects are influenced not only by variation of composition but also by electric fields [43,55–62]. For example, Figure 5.21 clearly illustrates that the photoinduced change of the dark discharge rate may successfully be locked in by applying an electric field $E \approx 3 \times 10^5\,\mathrm{V\,cm^{-1}}$ immediately after light exposure. For comparison, see curves 5 (Figure 5.21A) and 6 (Figure 5.21B).

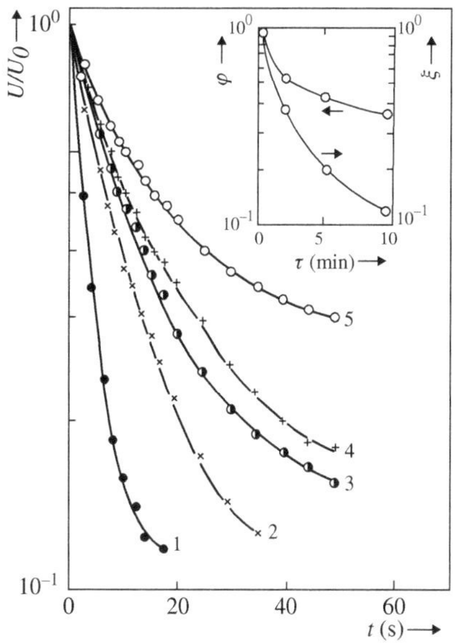

Figure 5.20 Dark discharge rate varying with a-$As_{0.15}Se_{0.85}$ sample dark-resting time and corresponding relaxation functions $\phi(\tau)$, $\zeta(\tau)$ (inset). (1) $\tau = 0$, (2) 2, (3) 5, (4) 10, (5) 90 min.

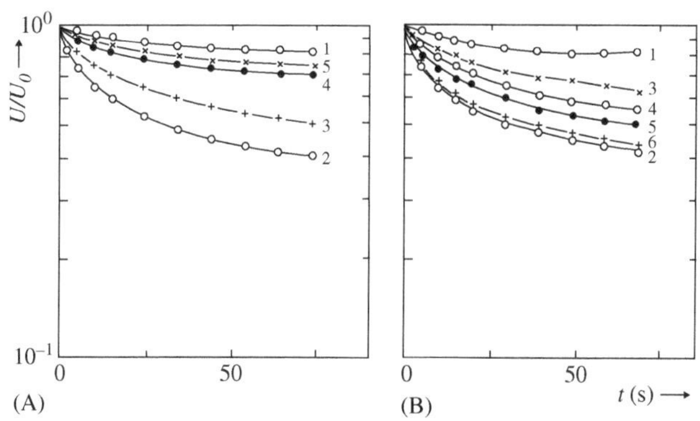

Figure 5.21 Surface potential dark decay in a-$As_{0.1}Se_{0.9}$ films ($d = 10\,\mu m$). (A) (1) Dark-rested sample, (2) exposed, (3)–(5) exposed and then dark-rested for 5, 20, and 30 min, respectively. (B) (1) Dark-rested sample, (2) exposed, (3)–(6) exposed and dark-rested for 30 min in the presence of a surface potential of 115, 170, 215, and 520 V, respectively [43,55].

Dark decay curves of the surface potential in double-logarithmic representation display two distinct rate processes $dU/dt\ t^{-(1-K)}$ and $dU/dt\ t^{-(1+K)}$ with crossover from one regime to the other at $t = t_d$. Detailed analysis of the discharge process shows that t_d shifts to smaller values with increasing As content and after exposure. Note that under low-charging voltages U_0, the surface potential at time t_d is approximately half the initial value and that t_d increases with U_0. The features observed in the dark decay study of As_xSe_{1-x} alloy films can be completely accounted for by a depletion-discharge model [25–33,43,44]. So the bulk process driving dark decay is emission and the sweep-out of holes from states near mid-gap to progressive

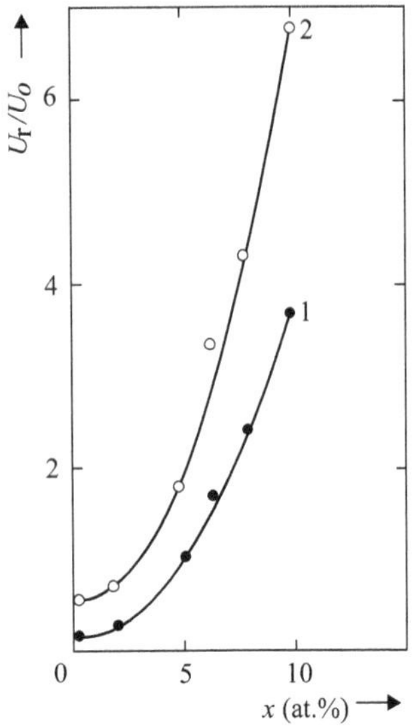

Figure 5.22 Dependence on residual voltage after (1) one and (2) five complete xerographic cycles of atomic percentage of As content in amorphous As_xSe_{1-x}.

formation of negative space-charge. The rise in dark discharge rate and the shift of t_d with arsenic concentration and photoexcitation may be due to enhanced thermal generation of holes from deep centers. From the temperature dependence of t_d, it is estimated that the emitting sites are located at 0.8–0.9 eV above the valence band mobility edge (Figure 5.22).

5.6 Residual Voltage in Se-Rich Photoreceptors

For the films under examination, the residual voltage U_r increases with exposure. For example, previous photoexcitation leads to an increase in U_{r1} (first-cycle residual voltage) for Se up to 3.7V caused by a decrease in $\mu\tau_L$ to $1.3 \times 10^{-7}\,cm^2V^{-1}$. Taking into account the invariance of $\mu^e = 4.9 \times 10^{-3}\,cm^2V^{-1}s^{-1}$ with light exposure, it is obvious that the lifetime reduction (from 5.4×10^{-5} to 2.6×10^{-5} s) is the only reason for photoinduced change.

During the continuous repetition of xerographic cycles, the residual voltage U_{rn} at the nth cycle was found to have the typical behavior shown in Figure 5.23.

The repetition of the xerographic cycle leads to the saturation of the residual potential (see Figure 5.23). The saturated value $U_{r\infty}$ corresponds to all the deep traps in the bulk being filled. In pure and Se-rich alloys of amorphous Se, this occurs typically after ~20–100 xerographic cycles. One can easily check the trap-filling interpretation if $U_{r\infty}$ shows any dependence on the cycling frequency to exclude a dynamic-equilibrium condition.

The saturation residual potential U_{rs} (or $U_{r\infty}$) provides a measure of the density of uniformly trapped carriers [18,25]: $U_{rs} = Ned^2/2\varepsilon$, where ε is the dielectric constant. An increase in U_{rs} in pre-exposed films indicates photoenhanced accumulation

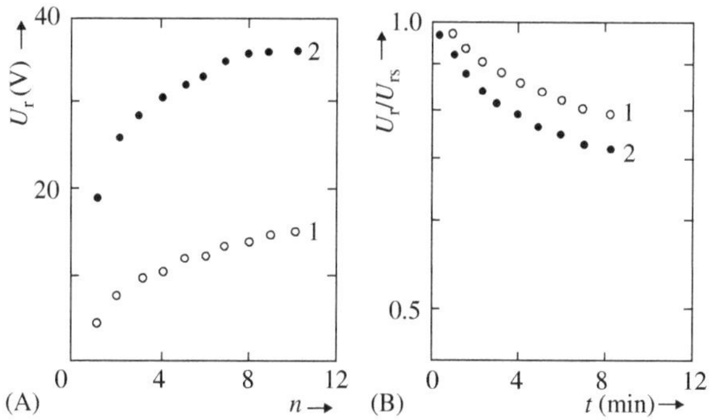

Figure 5.23 The buildup in the residual voltage with (A) number of xerographic cycles and (B) isothermal room-temperature relaxation of saturated residual voltage in (1) dark-rested and (2) photoexcited a-$As_{0.02}Se_{0.98}$ films.

of charge at deep centers. We obtain $N \approx 10^{15}\,cm^{-3}$ and $N^* \approx 4 \times 10^{15}\,cm^{-3}$ for dark-rested and pre-exposed $As_{0.1}Se_{0.9}$ films, respectively [43–55].

The residual potential decays to zero. This process is controlled by the spectrum of trap-release times. The trap energies can be deduced from an analysis of isothermal residual potential decay curves using $U_{rs} = \sum C_i \exp(-t/\tau_i)$, where $\tau_i^{-1} = \nu_i \exp[-E_i/kT]$ is the release time from the ith trap, ν_i is the frequency factor, and E_i is the trap depth. We find that deep levels in amorphous selenium reside at $E_i^h = 0.85\,eV$ and $E_i^e = 1.0\,eV$ for holes and electrons, respectively. Their depth becomes somewhat shallower with the addition of As. For example, $E_i^h = 0.80\,eV$, and $E_i^e = 0.90\,eV$ in $As_{0.1}Se_{0.9}$. The more rapid decay of the residual voltage U_{rs} in pre-illuminated films relative to dark-rested film (Figure 5.23B) indicates a slight decrease in the depth of those states. A comparison of room-temperature recovery in TOF and xerographic measurements demonstrates that relaxation of deep centers in pre-illuminated (exposed) films occurs on the same time scale. In other words, the deep gap centers which control the xerographic dark decay and residual voltage are, like the trapping centers discussed in TOF experiments, characteristically metastable.

Band-gap light can, in principle, have two distinct effects on the electronic structure of the mobility gap. It can either introduce new localized states or initiate conversion of traps of small cross-sections to traps of larger cross-sections [46]. The latter become accessible in deep level spectroscopy only after irradiation. So we may also consider such converted localized states as new localized states (created by irradiation).

5.7 Conclusion

Shallow and deep defect states in the band-gaps of Se-rich amorphous materials control technologically important photophysical parameters, such as drift mobility and deep trapping time (lifetime). Based on these experimental results, recommendations can be made about the best way for these parameters to be improved and the technical applications of Se-based amorphous materials are put in perspective.

5.8 References

1. R.A. Street, Appl. Phys. Lett. 41 (1982) 1060.
2. R.A. Street, J. Zesch, M.I. Thompson, Appl. Phys. Lett. 43 (1983) 1425.
3. R.A. Street, J. Zesch, M.I. Thompson, Appl. Phys. Lett. 43 (1983) 672.
4. R.A. Street, C.C. Tsai, M. Stutzman, J. Kakalios, Philos. Mag. B 56 (1987) 389.
5. W.E. Spear, W.E. Steemers, H. Mannsperger, Philos. Mag. B 48 (1983) L49.
6. E.I. Adirovich, Sov. Solid State Phys. 3 (1961) 2048 (in Russian).
7. P. Braunlich, P. Kelly, J.P. Fillard, Thermally Stimulated Relaxation in Solids (Top. Appl. Phys. 37) (ed. P. Braunlich) (Springer, Berlin, 1979).
8. Y. Gorokhovatsky, G. Bordovskij, Thermally Activational Current Spectroscopy of High-Resistance Semiconductors and Dielectrics (Nauka, Moscow, 1991) (in Russian).
9. R.A. Street, A.D. Yoffe, Thin Solid Films 11 (1972) 161.
10. B.T. Kolomiets, V.M. Lyubin, V.L. Averjanov, Mater. Res. Bull. 5 (1970) 655.
11. T. Botila, H.K. Henish, Phys. Status Solidi (a) 36 (1976) 331.
12. S.C. Agarwal, Phys. Rev. B 10 (1974) 4340.
13. S.C. Agarwal, H. Fritzsche, Phys. Rev. B 10 (1974) 4351.
14. P. Muller, Phys. Status Solidi (a) 67 (1981) 11.
15. A.A. Kikineshi, V.I. Mikla, I.P. Mikhalko, Sov. Phys. Semicond. 11 (1977) 1010.
16. V.I. Mikla, I.P. Mikhalko, Y. Nagy, J. Phys. Condens. Matter 6 (1994) 8269.
17. R.M. Schaffert, Electrophotography. Society of Photographic Scientists and Engineers (Focal Press, London, 1975), Chapters 1 and 2.
18. S.O. Kasap, in Handbook of Imaging Materials (ed. A.S. Diamond), (Marcel Dekker Inc., New York/Hong Kong, 1991), 329.
19. S.M. Vaezi-Nejad, C. Juhasz, Int. J. Electron. 67 (1989) 437.
20. S.M. Vaezi-Nejad, Int. J. Electron. 62 (1987) 361.
21. M. Abkowitz, R.C. Enck, Phys. Rev. B 25 (1982) 2567.
22. M. Abkowitz, R.C. Enck, Phys. Rev. B 27 (1983) 7402.
23. K.K. Kanazawa, I.P. Batra, J. Appl. Phys. 43 (1972) 1845.
24. S.W. Ing, Jr., J.H. Neyhart, J. Appl. Phys. 43 (1972) 2670.
25. M. Abkowitz, S. Maitra, J. Appl. Phys. 61 (1987) 1038.
26. A.R. Melnyk, J. Non-Cryst. Solids 35–36 (1980) 837.
27. M. Abkowitz, G.M.T. Foley, J.M. Markovics, A.C. Palumbo, Appl. Phys. Lett. 46 (1985) 393.
28. M. Baxendale, C. Juhasz, SPIE Proc. (1990) 1253.
29. S.O. Kasap, M. Baxendale, C. Juhasz, IEEE Trans. Indust. Appl. 27 (1991) 620.
30. S.O. Kasap, J. Electrostat. 22 (1989) 69.
31. M. Abkowitz, F. Jansen, A.R. Melnyk, Phil. Mag. B 51 (1985) 405.
32. M. Abkowitz, J. Non-Cryst. Solids 66 (1984) 315.
33. M. Abkowitz, J.M. Markovics, Solid State Commun. 44 (1982) 1431.
34. H. Scher, E.W. Montroll, Phys. Rev. B 12 (1974) 2455.
35. V.I. Mikla, I.P. Mikhalko, Y.Y. Nagy, A.V. Mateleshko, J. Mater. Sci. 35 (2001) 4907.
36. V.I. Mikla et al., Mater. Sci. Eng. B64 (1999) 1.
37. M. Abkowitz, Polym. Eng. Sci. 24 (1984) 1140.
38. M. Abkowitz, J. Non-Cryst. Solids 97–98 (1987) 1163.
39. M. Abkowitz, Physics of Disordered Materials (eds. D. Adler, H. Fritzsche, S.R. Ovshinsky), (Plenum Press, 1984, 483), New York.
40. A. Onozuka, O. Oda, I. Tsuboya, Thin Solid Films 149 (1987) 9.

41. S.O. Kasap, J. Phys. D 25 (1992) 83.
42. P.J. Warter, Appl. Optic. Suppl. 3 (1969) 65.
43. V.I. Mikla, D.G. Semak, A.V. Mateleshko, A.A. Baganich, Phys. Status Solidi (a) 117 (1990) 241.
44. V.I. Mikla, D.G. Semak, A.V. Mateleshko, A.R. Levkulich, Phys. Tech. Semicond. 23 (1989) 131 (in Russian).
45. S.O. Kasap, Handbook of Imaging Materials, (eds. A.S. Diamond and D.S. Weiss) 2nd ed. (Marcel Dekker Inc., New York, NY, 2002).
46. B. Polischuk, S.O. Kasap, V. Aiyah, A. Baillie, M.A. Abkowitz, Can. J. Phys. 69 (1991) 361.
47. O. Imagava, M. Iwanishi, S. Yokoyama, J. Appl. Phys. 60 (1986) 3176.
48. K. Tanaka, Curr. Opin. Solid State Mater. Sci. 1 (1996) 567.
49. V.M. Lyubin, Physics of Disordered Materials (eds. D. Adler, M. Kastner, H. Fritzsche), (New York, Plenum Press, 1985) 673.
50. K. Tanaka, Phys. Status Solidi B 246 (2009) 1744.
51. A.V. Kolobov, Photo-Induced Metastability in Amorphous Semiconductors (Wiley-VCH, 2003), Berlin.
52. V.I. Mikla, V.V. Mikla. Metastable States in Amorphous Chalcogenide Semiconductors (Springer, Heidelberg), 2009
53. K. Tanaka, Chalcogenide glasses, in Encyclopedia of Materials: Science and Technology, Elsevier Science Ltd. (2001) 1123–1131.
54. V.I. Mikla, V.V. Mikla, Trap Level Spectroscopy in Amorphous Chalcogenide Semiconductors (Elsevier Insights, New York, NY, 2010).
55. V.I. Mikla, J. Phys. Condens. Matter 8 (1996) 429.
56. V.I. Mikla et al., 14th General Conf. GCMD-14 (Madrid, 1994), 1055.
57. V.I. Mikla et al., 15th General Conf. Condensed Matter Division (Baveno-Stresa, Lago Maggiore, Italy, 1996), 20, 103.
58. V.I. Mikla, Int. Workshop on Adv. Technol. Multicomponent Solid Films (Uzhgorod, 1996), Book of Abstracts, 84.
59. V.I. Mikla et al., Patent USSR 4088420 (1986).
60. V.I. Mikla et al., Patent USSR 4273793 (1987).
61. V.I. Mikla et al., Patent USSR 4638891 (1988).
62. V.I. Mikla, I.P. Mikhalko, Y.Y. Nagy, J. Non-Cryst. Solids 142 (2001) 1358.

6 Effect of Antimony Alloying on Photoelectronic Properties of a-Se

Amorphous chalcogenide semiconductors, which do not contain such conventional additives as arsenic, have recently interested researchers due to their potential application in X-ray digital sensors. Reasonably, we consider fundamental photoelectronic properties of the Sb_xSe_{1-x} ($0 \leq x \leq 0.10$) system. Various conventional and less familiar techniques, such as TOF transient photoconductivity and xerographic measurements, are used to examine properties of these promising but poorly studied noncrystalline materials.

Glasses and amorphous films are fabricated by melt-quenching and vacuum deposition onto unheated substrates, respectively. The most essential (in the context of technical application) results may be summarized as follows: it is observed that the addition of antimony increases the DC conductivity, decreases the electron mobility, changes the xerographic parameters, and decreases the optical gap of the system. TOF experiments indicate that electron transport in a-Sb_xSe_{1-x} films is controlled by a set of shallow traps located at ~0.33 eV below E_c, whose concentration increases with Sb addition. Xerographic dark discharge measurements showed that the decay of the surface potential is essentially due to bulk thermal generation of electrons and their subsequent sweep-out and depletion. Electron emission occurs from mid-gap localized states. When a-Se is alloyed with antimony, the dark discharge becomes more rapid due to an increase in the volume density of the mid-gap electron emission centers with antimony concentration. Cycled-up xerographic residual voltage measurements indicate that the saturated residual voltage increases with Sb addition and is caused by an increase in the concentration of electron deep traps ($N_t \sim 10^{15}\,cm^{-3}$). The xerographic photosensitivity for Sb_xSe_{1-x} alloys is greater at longer wavelengths, $\lambda > 670\,nm$, than for pure Se. The compositional dependence of the electronic properties is related to incorporation of Sb atoms in a glassy selenium matrix.

In general, glasses are often preferred over crystalline compounds with similar characteristics because of favorable physical properties (including mechanical properties) and good processability. In addition, a lack of translational regularity in atom positions allows researchers to tailor physical properties to specific applications by adjusting chemical composition.

Several kinds of applications of amorphous chalcogenides are commercially available, practically utilized, or at the stage of technical realization [1–5].

Amorphous Chalcogenides. DOI: 10.1016/B978-0-12-388429-9.00006-6

Non-oxide chalcogenides have been extensively studied due to their interesting electronic and optoelectronic properties [6,7]. There is, at present, an increased interest in exploring As-free Se-based alloys, for example, those that contain Sb. Antimony is a group III element that has a high Z and acts in principle as a cross-linking agent between Se chains, thereby providing thermal stability.

Sb_xSe_{1-x} glasses are promising candidates for applications where low melting temperatures, low thermal conductance, and high viscosity are required. Thin films from binary Sb–Se and ternary Sb-containing systems can undergo amorphous-to-crystalline phase transition under the influence of illumination and may be considered as reversible optical recording candidates.

Historically, selenium in the noncrystalline phase has a special attraction in semiconductor physics research because of its commercial importance as a xerographic photoreceptor (see Chapter 4 for more details).

As reported in recent conferences on noncrystalline solids, a-Se photoconductor-based flat panel X-ray detectors are now at a stage that can provide excellent images, and they are being employed in various X-ray imaging applications—both medical and nonmedical [8].

There are several important factors that make Se a particularly suitable testing object for the study of impurities and/or alloying effects on the electrical properties. Along with its elemental nature, its relative structural simplicity, ambipolarity, and high photosensitivity should be mentioned. The alloying effects in a-Se have been well documented and reviewed for such additives such as As, S, Te, and P [2,9–11]. At the same time, there is only limited information about the basic properties of the amorphous Sb–Se system [12–19]. Recently, thin films of these materials have become of interest for use in optical data storage [20]. Next, the effect of antimony on the photoelectronic properties (especially on the deep and shallow states) of selenium is considered. These states in the band-gap are shallow and deep. They preferentially determine photoelectronic properties of noncrystalline materials and control the drift mobility, dark decay, and cycled-up residual surface potential on capacitively charged specimen films. They also act as trapping or thermal emission centers.

6.1 Preparation of a-Sb_xSe_{1-x} and Measurement Techniques

Glassy Sb_xSe_{1-x} alloys ($x < 0.05$) were prepared from the pure elements. Cleaned silica tubes containing a mixture of the appropriate amount of constituents Sb and Se were evacuated to 10^{-5} torr and sealed. The contents of the tubes were melted in a furnace and continuously agitated for 10 h to ensure good homogeneity. Following heating, the ampoules were quickly quenched in cold water from 800°C to a temperature of 0°C.

Amorphous film samples from quenched materials were prepared by conventional vacuum deposition on pre-cleaned, etched, and pre-oxidized Al substrates: these were cleaned in neutral detergent, deionized water, ethyl alcohol, and acetone and then oxidized in air at 1,800°C before deposition. Deposition of the

Sb_xSe_{1-x} alloys was performed at a substrate temperature close to room temperature. It should be noticed that while the substrate is aluminum, the layer directly under the film is aluminum oxide. Care was taken to avoid Sb fractionation. Excluding the initial and final stages of the evaporation process by shuttering, the deposited antimony content is relatively uniform across the film thickness, as can be seen from our electron probe microanalysis data, and it is close (± 0.5 at.%) to the starting source value. A typical coating rate was 1–2 $\mu m\, min^{-1}$. The film thicknesses ranged from 10 to 20 μm. Prior to measurement, the Sb_xSe_{1-x} films prepared were aged in the dark under normal laboratory conditions (i.e., $T \approx 293$ K and relative humidity 70–80%) for 2–3 weeks to allow for their physical properties to equilibrate.

Sb alloying effects on thermal generation and both deep and shallow trapping of carriers have been examined using xerographic and TOF techniques [2,21,22]. In the xerographic measurement, the sample is charged to a potential V_0 by passing it under a corotron (corona charging) device that deposits charges of the appropriate sign on the surface of the film. The surface potential is then measured by a transparent probe and an electrostatic voltmeter. The photoreceptor, the corona-charging units, and a transparent probe were housed in a well-shielded and dark environment. Following the initial-charging process, the sample is exposed to strongly absorbed 450 nm step illumination from a tungsten light source during which time the decay of the surface electrostatic potential is monitored by the electrostatic voltmeter. The surface potential decays to a potential V_r (termed the residual potential), and the resulting PID curve can be used to determine the xerographic photosensitivity of the sample. This xerographic step could be repeated any number of times to obtain a cycled-up residual surface potential V_{rn} as a function of the xerographic cycle n.

The xerographic measurements have been complemented by TOF transient photoconductivity measurements to study the charge transport properties of Sb_xSe_{1-x} alloys. The experimental TOF technique has been described in the literature by a number of authors [2,21,22]. In a-Se-based materials, it is more useful to measure time-resolved transits in the current mode of operation because of the dispersion and relatively long transit at low fields.

In the present case, we used a nitrogen gas laser to photoexcite the carriers. The nitrogen gas laser provided a short light pulse of duration $\sim$10 ns at a wavelength of 337 nm. A semitransparent gold electrode was sputtered onto the surfaces of the films as the top electrode. The photocarriers are generated at the sample surface due to the high absorption coefficient of Sb_xSe_{1-x} glasses at a wavelength 337 nm ($\alpha > 10^4\, cm^{-1}$), so the species of drifting carriers can be chosen by changing the polarity of applied voltage across the sample. The transient current was amplified and displayed on a single-event storage oscilloscope. The TOF measurement was carried out under the single-shot mode of operation, and between each measurement, the sample was short-circuited and stored (relaxed) in the dark to allow any bulk space-charge buildup to decay. Small-signal conditions were maintained throughout all the measurements.

6.2 Fundamental Properties

The Sb–Se glasses, like many other chalcogenide glasses, are expected to show glass transition and crystallization behavior. This is caused by structural rearrangements and relaxation processes toward equilibrium. A proper description of thermal transformation of these glasses with a well-defined thermal history is crucial to understanding their basic properties. Usually, most of phase transformation studies (glass transition and crystallization) of chalcogenide glasses have been carried out by conventional differential scanning calorimetry (DSC) [23].

Temperature-modulated differential scanning calorimetry (TMDSC) is a relatively new technique [24] that has been found to be an accurate tool for measuring the temperatures (and enthalpy) of various transitions occurring in a material. The description and benefits of the TMDSC technique have been documented in several recent papers, e.g., [25]. The TMDSC signals (thermograms), as modulated heat flow (MHF), deconvoluted heat flow (DHF), non-reversing heat flow (NHF), and reversing heat flow (RHF), are exemplified for composition a-Se 0.5% As in Figure 6.1.

The endo-peak and area for evaluation of ΔH_g are shown in Figure 6.1A. The enthalpy of crystallization ΔH_c, the crystallization onset, and maximum rate temperatures T_c and T_p were also extracted from NHF curves. In Figure 6.1B, one can see an example of a typical exo-peak and an area for evaluation of ΔH_c, T_c, and T_p. Figure 6.1A and B is for sample a-$Sb_{0.15}Se_{0.85}$.

The T_g was operationally defined as the temperature where the heat capacity versus temperature behavior evinces an inflection point through the glass transformation, as seen in Figure 6.1B.

Figure 6.2 illustrates the dependence of T_g on the composition of the Sb–Se glasses. The parameter T_g behaves in the following manner with progressive Sb addition: it increases with Sb addition (up to 1 at.%), reaches a local maximum, decreases to 10 at.%, raises up to 15 at.% and then decreases. These results obtained for binary Sb–Se glasses agree qualitatively with similar effects when adding Sb to ternary chalcogenide glasses.

It can be noted that the compositional dependence of ΔH_g also has a maximum at compositions of 1 at.% Sb and at 15 at.% Sb. In addition, the dependence of the specific heat capacity C_p and the change in the specific heat capacity ΔC_p on the composition of Sb–Se glass alloys both show the same maximum at 1 at.% Sb and some peculiarity at 15 at.% Sb.

The crystallization process follows the glass transition, and the rate of crystallization depends on the viscosity of the liquid. In comparing the crystallization behavior of glasses with limited stability (and which may crystallize even spontaneously), it is more informative to determine the crystallization and glass-transition temperature difference rather than the absolute values of T_g and T_c [15]. Generally, a wider difference T_c–T_g is directly associated with higher kinetic resistance to crystallization [23]. The plot of T_c–T_g versus Sb content in Figure 6.3 shows that the thermal stability is at a maximum when the Sb content reaches 1 at.%. Therefore, it is not surprising that the dependence of the crystallization maximum rate temperature T_p and the difference between T_c and T_g on the composition of Sb–Se glass alloys have similar

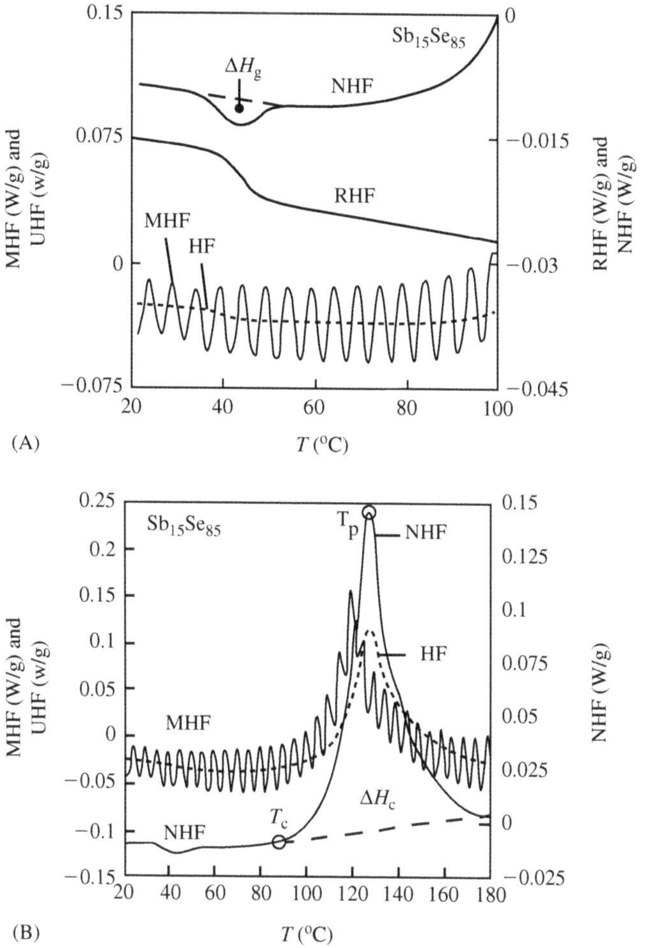

Figure 6.1 Typical TMDSC signal with underlyinging total HF and deconvoluted RHF and NHF; (A) shows the area for the evaluation of the relaxation enthalpy ΔH_g; (B) shows the area for the evaluation of the crystallization enthalpy ΔH_c, crystallization onset T_c, and peak crystallization rate T_p for temperatures on the NHF curve [14].

shapes on the T_g curves in Figure 6.2. According to these results (ΔH_g, T_p, and T_c–T_g), the most stable glasses are expected to be around 1 at.% Sb and 15 at.% Sb [14].

These precise measurements show that the overall change in T_g observed upon alloying up to 25 at.% Sb in Se is 10°C only [14]. At the same time, addition of 25 at.% As in Se produces an essential change in T_g of about 80°C.

Although Sb and As are both group V elements, their interactions evidently differ when alloyed with Se. We agree with Tonchev and Kasap [14] that the different size of Sb and As atoms may be more important than their periodic table valence. Reasonably, it is easy to obtain amorphous As_2Se_3 glasses and difficult to obtain Sb_2Se_3 glasses.

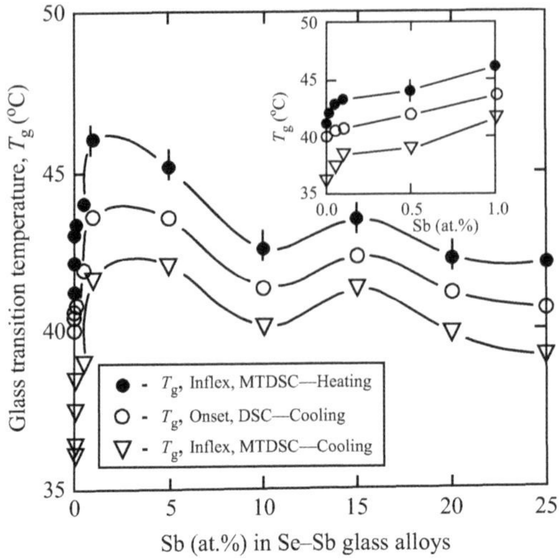

Figure 6.2 Dependence of the T_g on the composition of Sb–Se glass from TMDSC and DSC (heating and cooling scans). Inset: the same dependence for the low concentration shown [14].

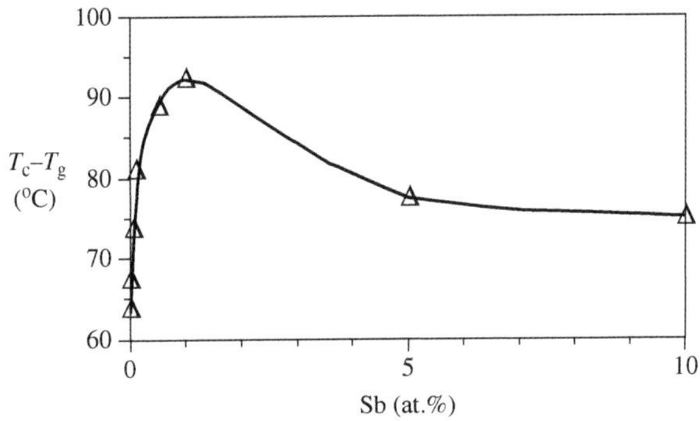

Figure 6.3 Dependence of T_c–T_g on the composition of the Se–Sb glass alloys [15].

In addition to thermodynamic parameters, microhardness, density, DC conductivity, optical band-gap, main extremum at the stoichiometric composition $Sb_{0.40}Se_{0.60}$, and other important structural-sensitive physical parameters show similar compositional trends with local extrema. These are observed at around the same composition ~1 at.% [26,27]. This phenomenon is inherent also to other Se-based alloys and will be explained in a separate section on the basis of the Phillips–Thorpe constraint theory [28–32]. Here, we mentioned only one possible interpretation of the results: initially, up to 1 at.% Sb enters the Se backbone and provides cross-linking of Se chains.

For all Sb–Se compositions used in this study, the dark DC conductivity can be expressed by a conventional, Arrhenius-type, relation

$$\sigma = \sigma_o \exp(-E_\sigma/kT), \tag{6.1}$$

Table 6.1 Room-Temperature DC Conductivity σ_{ROOM}, Activation Energy E_σ, Dark Discharge Rate dV/dt, Time Required to Discharge the Films to Half of Their Initial-Charging Voltage (V_0) $t_{V0}/2$, and the Optical (Tauc) Gap E_0 of a-Sb_xSe_{1-x} [26]

Sb content (at.%)	σ_{ROOM} (Ω^{-1} cm^{-1})	E_σ (eV)	dV/dt (V^{-1}s^{-1})	$t_{V0}/2$ (s)	E_0 (eV)
0	1×10^{-14}	1.05	7.8×10^{-2}	267	2.04
2	1×10^{-13}	1.02	2.0×10^{-1}	81	2.02
5	8×10^{-11}	0.90	1.83	76	
8	9×10^{-10}	0.84			
10	4×10^{-10}	0.76			1.85
20	1×10^{-9}	0.72			1.62
40	2×10^{-8}	0.62			1.50

where E_σ is the activation energy and σ_0 is the conductivity pre-exponential factor. These electrical and main fundamental parameters measure 300 K and are presented in Table 6.1.

It is evident that DC conductivity (σ_{ROOM}) and dark discharge rate (dV/dt) increase with Sb content. In contrast, activation energy (E_σ) of DC conductivity, characteristic time of dark discharge rate ($t_{V0}/2$), and optical gap decrease with increasing antimony content in the Sb–Se system [26].

The glassy system Sb–Se optical absorption edge (in the Urbach tail region) shifts to longer wavelengths, and the slope becomes smaller as the percentage of Sb increases. This is a general trend in the composition range examined. Absorption above the fundamental edge follows the familiar Tauc law characteristic of the most of amorphous chalcogenides

$$\alpha h\nu = C(h\nu - E_T)^2, \tag{6.2}$$

where $h\nu$ is a photon energy, E_T is the Tauc gap, and C is a constant indicating how steeply the absorption rises with energy. For most compositions, there was no deviation from the square law even at the highest absorption value measured in this study; exceptions were for pure a-Se and films with less than 1 at.% Sb content. These latter follow the linear law

$$\alpha h\nu = C_1(h\nu - E_1) \tag{6.3}$$

in the range 2.1/2.5 eV. Here, C_1 is a constant and E_1 is an extrapolated optical gap. The addition of less than 1 at.% Sb to pure Se is enough to cause a complete transition from the anomalous linear behavior of Tauc law. The room-temperature values of E_T (averaged over a series of the samples) for various Sb_xSe_{1-x} compositions are given in Table 6.1. It should be noted that the Tauc gap E_T is close to that of pure Se for small Sb concentrations (<1 at.% Sb), and then it decreases almost linearly up to nearly 40 at.% Sb.

6.3 Dark Discharge

Typical dark discharge characteristics for pure Se and Sb_xSe_{1-x} photoreceptors are shown in Figure 6.4. It is apparent that for pure a-Se, the decay of the surface potential is relatively slow. Comparison of the respective characteristics for a-Sb_xSe_{1-x} with the dark discharge behavior of pure a-Se clearly shows that alloying a-Se with antimony increases the dark decay rate. The discharge rates in a-Sb_xSe_{1-x} were not constant—they changed with time after charging.

There are several physical processes that may lead to the decay of the surface potential. The mechanism of the dark decay in a-Se-based films and the presently accepted model of this phenomenon are explained in detail in Chapter 5. The physical processes that may cause the dark decay of surface potential [2,33–39] include surface generation, substrate injection, bulk thermal generation of carriers of one sign, injection of trapped electrons and their consequent transport across the sample, and depletion.

With relatively thick films (L = 10–50 μm) and a good blocking contact between films and the preoxidized Al substrate, the dark decay dominates in amorphous Sb_xSe_{1-x} layers.

In a series of experiments carried out on a composition series of glassy Sb_xSe_{1-x} alloys, it was found that the time-dependent dark decay rate of the potential to which a dark relaxed film has been charged is controlled by the depletion discharge process. In general, the xerographic depletion discharge model is based on bulk thermal generation involving the ionization of a deep mobility gap center to produce a mobile charge carrier of the same sign as the surface charge and an oppositely charged ionic center (see Chapter 5). Assuming negative charging for amorphous Sb_xSe_{1-x} layers deposited onto unheated substrates, a mobile electron would be thermally generated, and the ionized center would be positive. As thermally generated

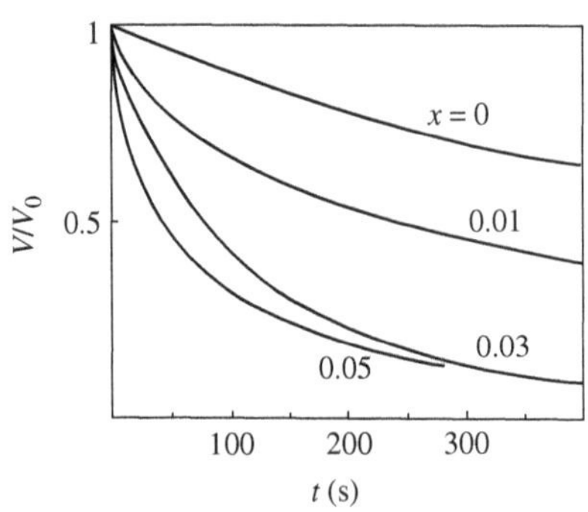

Figure 6.4 Dark discharge of surface potential on a-Sb_xSe_{1-x} layers.

holes are swept out by the electric field, a positive bulk space-charge builds up with time in the specimen, causing the surface potential to decay with time.

The spatial distribution of bulk trapped charges can be determined from analysis of XTOF measurements [40,41]. With the emergence of nonimpact printers for electronically processed or stored information, this less familiar technique has been found to be very useful as a nondestructive method of investigating transport properties—in our case, the spatial distribution of bulk space-charge of the photoreceptors used in these systems. The technique is a xerographic version of conventional TOF and can be employed in parallel with traditional measurements for photoreceptor characterization [40,42].

In order to better characterize the state of Se-like photoreceptors, detailed knowledge of the electric field's spatial distribution within a photoreceptor and its cyclic variation are required. Typically, this technique has been utilized on samples that have electrodes on opposite sides of the layer and that are limited to relatively low electric fields by discharge at discrete, low-field breakdown regions. XTOF is a technique without exposed surface electrodes, which is used as an adjunct measurement during xerographic cycling of a photoreceptor under machinelike or special diagnostic conditions. Those discrete regions limiting the electric field in the electrode case are not important in the XTOF case since very little transverse charge movement occurs.

XTOF was successfully carried out at the IBM research laboratories [43] and more recently at the Xerox laboratories [44]. The principle of operation of the technique has been described here, and it is illustrated in Figure 6.5.

In the XTOF experiments, a sheet of photoexcited carriers at the sample film's top surface drifts across the bulk under the combined action of the uniform field established by a corona charge and the local field produced by any trapped charge. Use in the small signal facilitates analysis and moderates any influence of measurement upon the photoreceptor's cyclic properties. The temporal voltage change is $V(t) = n\mu F(x)$, where n is the photoinduced mobile charge, μ is the drift mobility, and $F(x)$ is the electric field measured from the exposed surface. Through

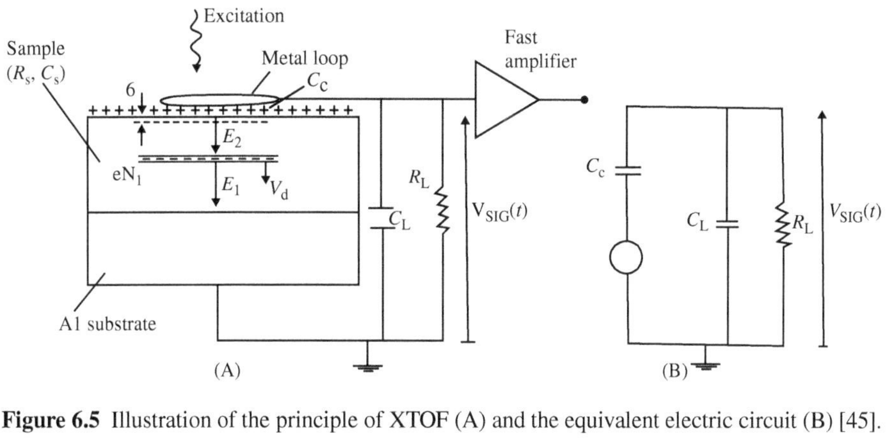

Figure 6.5 Illustration of the principle of XTOF (A) and the equivalent electric circuit (B) [45].

transformation of the data (including that of the surface potential), the temporally dependent signal is connected to a spatial representation of the electric field along the direction of charge flow normal to the layer plane.

For the films under examination, all transit pulses observed in the space-charge free sample are rectangular with a Gaussian tail. Pulse shapes exhibit a characteristic distortion when there is a trapped homocharge. Here, we note only that if the drift mobility is independent of position, then the magnitude of the XTOF waveform at any time is proportional to the local electric field at the instantaneous position of the drifting centroid charge. There is a transformation from the derivative of surface voltage versus time to internal local electric field versus position. Finally, the spatial derivative of the local electric field gives the space-charge distribution from Poisson's equation.

If dark decay is caused by depletion discharge, a positive bulk space-charge builds up with time in amorphous Sb_xSe_{1-x} layers deposited onto unheated substrates. In fact, XTOF signal distortion may be considered as a strong argument in favor of the presence of bulk space-charge in the specimens.

Figure 6.6 depicts typical dark discharge data for four different compositions, which illustrate the predicted characteristics of depletion discharge behavior. Inflections in the log–log plots at the respective depletion times (marked by an arrow for the case of $Sb_{0.03}Se_{0.97}$ composition) are readily identifiable. From the temperature dependence of depletion time, it is estimated that the emitting sites are located $\sim 0.9 \pm 0.05\,eV$ below the conduction band mobility edge.

Some special problems can, however, complicate the observation of a depletion kink in pure a-Se. The dark discharge rate was typically so slow in a-Se that results were always perturbed by injection. The central reasons for pure Se possessing good dark decay characteristics are as follows:

- The number of deep localized states ($\sim 10^{13}\,cm^{-3}$) in the mobility gap is remarkably small and integrated.
- The energy location of these states is deep ($E_t \sim 1.0\,eV$) in the mobility gap so that the thermal generation process of carriers from these centers is slow.

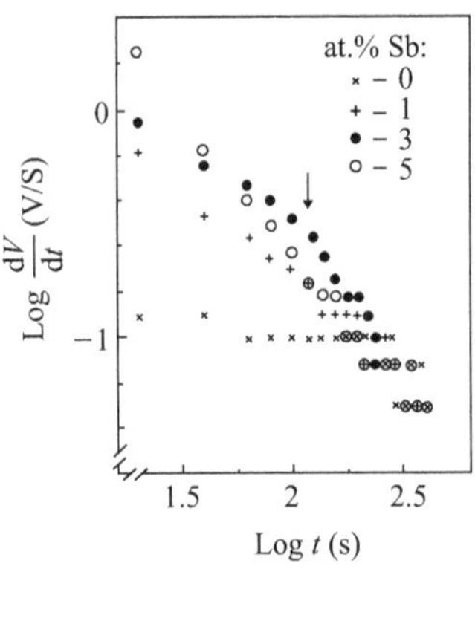

Figure 6.6 Typical dark discharge data for amorphous Sb–Se [26] (Sb concentration is shown in the top right).

It is found that in a-Sb_xSe_{1-x} films deposited on the substrates held at room temperature, electrons (the mobile carrier species) are depleted (n-type system) during dark decay, leaving behind a deeply trapped positive space-charge. Note that the same situation prevails in alkali-doped a-Se [46].

6.4 Transient Photoconductivity

During the 1960s and 1970s, charge transport and trapping were extensively studied. Although researchers at that time had a lack of proper understanding of the physics of amorphous semiconductors, the wealth of the experimental data was later put into perspective by various authors [2,47]. This increased interest seemed natural because the product of the charge carrier drift mobility and lifetime $\mu\tau$, called the carrier range, determines the xerographic performance of a photoreceptor.

In crystalline semiconductors, the most common technique for the measurement of carrier mobility involves the Hall effect. In contrast, in noncrystalline materials, experimental data is both fragmentary and anomalous [11,47]. Measured Hall mobility is typically of the order of 10^{-1}–$10^{-2} cm^2V^{-1}s^{-1}$ and is frequently found to exhibit an anomalous sign reversal with respect to other properties, which provides information concerning the dominant charge carrier. Thus, the Hall effect measurements are of minimal value in the study of macroscopic transport in amorphous semiconductors.

A second common technique in the case of crystalline semiconductors is the procedure developed by Haynes and Shockley [48] for the measurement of carrier drift mobility. The technique is restricted to the study of minority carrier transport. Excess packets of majority carriers rapidly spread out during transit so that information concerning their transit time along the specimen is lost. Since this spreading occurs in the order of dielectric relaxation time, it will take place more slowly in materials of low electrical conductivity, as is the case for a range of amorphous semiconductors of interest. These amorphous semiconductors are characterized by carrier drift mobility, which is typically much lower than that in crystalline semiconductors, so that transit times must be reduced to acceptable short values if the technique is to be of value (Figure 6.7).

The experimental configuration, developed by Spear [49,50], achieves the above objective and has been widely used in the study of carrier transport in low mobility materials. The sample is sandwiched between two electrodes. These are typically Au and Al, the former being semitransparent. Excess charge carriers of both signs are created in equal concentrations, close to the upper surface of a thin specimen film. A short-duration excitation pulse of strongly absorbed photons is normally used for this purpose. The external resistance R is much less than the sample resistance. Reasonably, the applied bias V appears across the thickness L of the specimen. Normally, the injected charge is much less than that on the electrodes. Thus, the electric field in the sample remains relatively unperturbed. Depending upon the polarity of the electric field applied across the specimen, carriers of one sign (e.g., electrons)

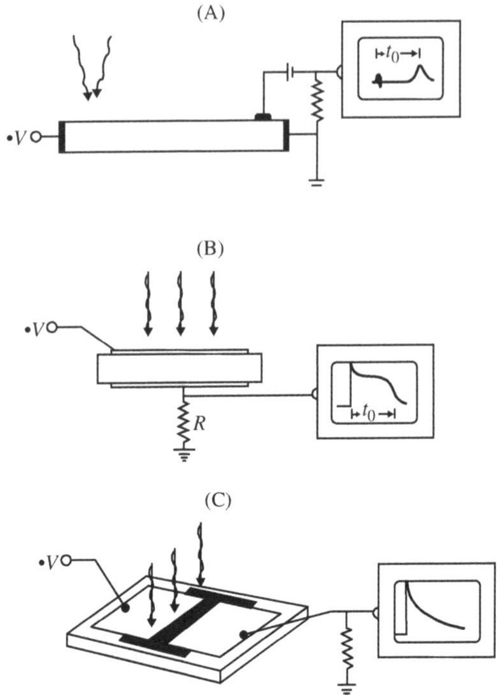

Figure 6.7 Various experimental techniques for the examination of carrier drift mobility. (A) The Haynes–Shockley technique, (B) the Spear technique, (C) the transient photodecay technique.

become neutralized nearly immediately after reaching the top electrode. At the same time, carriers of opposite polarity drift across the film, inducing a signal in the series resistor R (in other words, in the external circuit). Usually R is chosen to fulfill the following requirement: the circuit time constant $C_{sample}R$ must be shorter than the transit time t_T (so-called I, or current mode of operation—Figure 6.8). This signal will fall to zero when all carriers have completed their transit, allowing an identification of the mean carrier transit time, and thus the drift mobility. It is important to note that the carrier of either sign may, in principle, be examined since the dielectric relaxation time in suitable materials is much longer than the transit time through the thin film.

The principle of the TOF measurement is illustrated in Figure 6.9 [2].

Under the idealized conditions of an infinitely thin sheet of excess carriers drifting through the specimen with zero dispersion and with suitable blocking electrodes to prevent undesirable reinjection of charge, the voltage induced across the sampling resistor R would fall abruptly to zero at a time equal to the transit time of the carriers. In practice, of course, a spread of arrival times always occurs due to diffusion and other effects, so that a reasonably well-formed signal might be of the type shown in Figure 6.8B. With such a pulse shape, it remains comparatively easy to identify a mean transit time for the excess carriers, from which the average drift mobility may be computed.

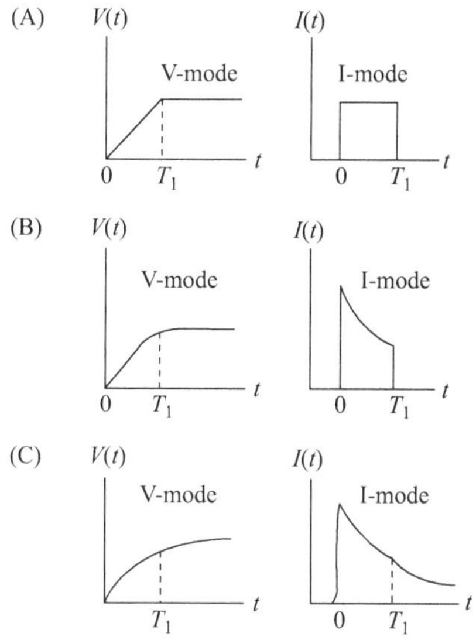

Figure 6.8 Transient current (*I* mode of operation) and voltage (*V* mode of operation) waveforms showing effect of trapping and detrapping. (A) No trapping; (B) trapping, no detrapping; (C) trapping and detrapping.

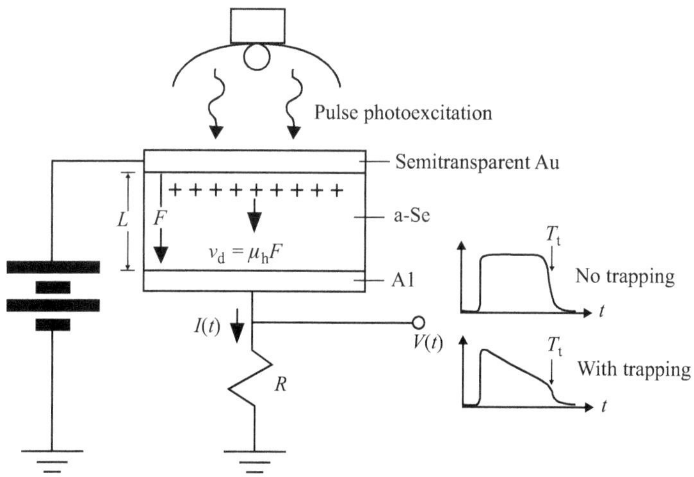

Figure 6.9 The principle of TOF experiments [2].

The advantage of the TOF measurement lies in its ability to monitor the motion of charge carriers across a photoreceptor film and to provide direct evidence as to whether the photoinjected carriers are making it across the sample.

Both the electron and the hole drift mobility can be measured in amorphous selenium by the TOF technique outlined previously. At temperatures above 200 K, a well-defined transit pulse is observed. The transient profile for a well-relaxed (dark-adapted) sample is a quasi-rectangular pulse. The photocurrent remains approximately constant up to t_T, and then decreases abruptly. The signals indicate essentially Gaussian transport. The transit time t_T was defined to correspond to the break point in the photocurrent. This represents the transit time of the fastest carriers. The mobility μ is then calculated using $\mu = L^2/(t_T V_0)$, where V_0 is the applied voltage and L is the sample thickness. These room-temperature values of the drift mobility, $\mu^h \approx 2 \times 10^{-1}\,cm^2V^{-1}s^{-1}$ and $\mu^e \approx 7 \times 10^{-3}\,cm^2V^{-1}s^{-1}$, are in remarkably good agreement with values reported in other sources [2,9,11,32,49].

It is important that for a-Sb_xSe_{1-x} films deposited on a room-temperature substrate, it was not possible to detect any pulses associated with the transit of hole carriers (Figure 6.10) [17,26].

In all samples, hole response showed a rapid featureless quasi-exponential decay with no apparent break.

There are two plausible explanations. First, the signal is limited by the presence of deep gap states with extremely high efficiency for carrier trapping. Second, it might be argued that alloying may cause an increase in the conductivity of the samples so that TOF experiments are no longer applicable. However, results showed that although the conductivity increases with Sb addition, it remains sufficiently low ($\leq 10^{-11}\,\Omega^{-1}cm^{-1}$) for TOF experiments to be applicable. It is interesting that hole response in As-alloyed a-Se has also been found to be undetectable in the range 2–4 at.% As [51,52].

In contrast, the effect of Sb on electron transport is slight. Although Sb alloying increases the transit time dispersion, the transit time shown contains a clearly identifiable break in the waveform. The electron drift mobility in a-Sb_xSe_{1-x} alloys exhibits Arrhenius behavior (Figure 6.11). The experimentally observed activation energy of a-Se, namely $E_\mu \approx 0.33 \pm 0.01\,eV$, remains almost insensitive to Sb addition. This

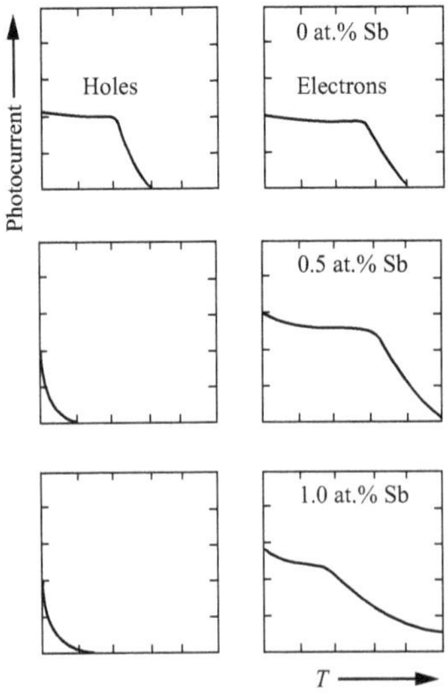

Figure 6.10 Oscilloscope traces of typical TOF transient current waveforms in amorphous selenium alloyed with Sb. For Se, $E = 8.0 \times 10^6\,Vm^{-1}$ and $0.1\,\mu s\,div^{-1}$ for holes and $2\,\mu s\,div^{-1}$ for electrons. For 0.5 at.% Sb, $E = 9.0 \times 10^6\,Vm^{-1}$ and $0.1\,\mu s\,div^{-1}$ for holes and $2\,\mu s\,div^{-1}$ for electrons. For 1 at.% Sb, $E = 8.8 \times 10^6\,Vm^{-1}$ and $0.1\,\mu s\,div^{-1}$ for holes and $5\,\mu s\,div^{-1}$ for electrons [17].

fact means that Sb addition must increase the total concentration of the shallow electron traps that control the drift mobility. At the same time, the mobility value decreases with increasing Sb content. It is reasonable, within the framework of a shallow-trap-controlled mobility model [47], to interpret our TOF observations in the following manner: in pure Se, electron transport is controlled by a narrow manifold of traps located about 0.33 eV from the conduction band mobility edge. The addition of Sb to Se, as we believe, broadens the distribution of shallow traps, thus increasing the relative dispersion of photocurrent transients. The decrease in electron mobility with increasing Sb content can be accounted for by the increase in the density of shallow traps with E_{μ} remaining constant at ~0.33 eV.

Other recent measurements [15] of the charge transport properties performed using conventional and IFTOF techniques on a-Sb_xSe_{1-x} films deposited on the substrates held at ~70°C indicate the following: Sb addition to a-Se affects the hole mobility only slightly whereas there is a substantial decrease in the electron mobility with the Sb content. An initial increase in both the electron and hole lifetimes with small amounts of Sb addition is observed. Both lifetimes reach a maximum at ~1 at.% Sb, beyond which they both decrease with Sb addition [15]. Note that the initial enhancement in the carrier lifetime exceeds the drop in the drift mobility with Sb content. This leads to enhanced electron and hole ranges, which are defined as the drift mobility (μ) and lifetime (τ) product, $\mu\tau$. This physical parameter determines the performance of many amorphous semiconductor devices: solar cells, photoreceptors, photodetectors, thin film transistors, etc. (Figure 6.12).

For charge carrier ranges in amorphous semiconductors to be determined, various measurement techniques are used. Among these we mentioned: conventional and XTOF transient photoconductivity, DFTOF, and IFTOF techniques. Although all or most $\mu\tau$ measurements have utilized TOF-type transient photoconductivity experiments [50,53–57] and the Hecht analysis [58], xerographic measurements

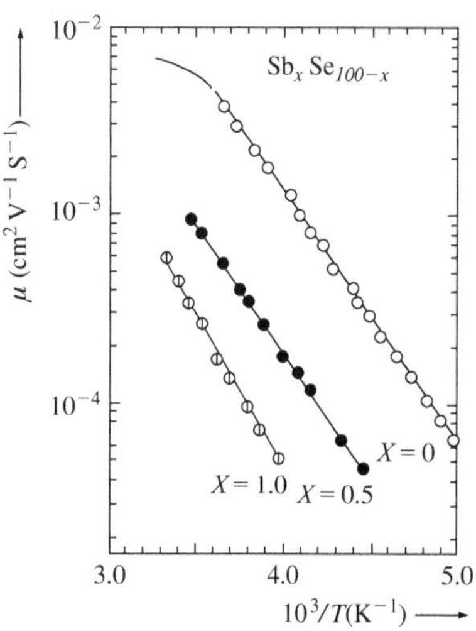

Figure 6.11 Temperature dependence of electron drift mobility in amorphous Sb_xSe_{100-x} deposited onto substrates held at room temperature.

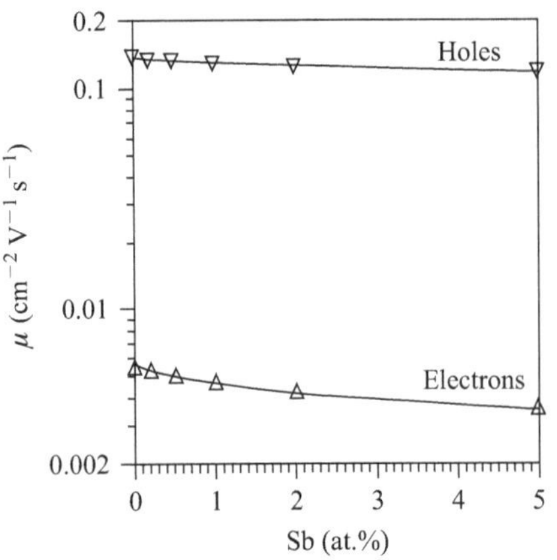

Figure 6.12 Dependence of the mobility of electrons and holes on the composition of Sb–Se films from TOF experiments [15]. Films were deposited onto the substrates held at ~70°C.

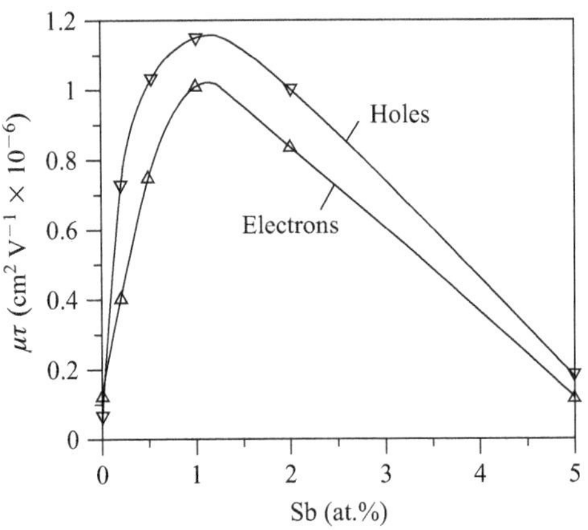

Figure 6.13 Dependence of the carrier range on the composition of Sb–Se films from TOF/IFTOF experiments [15]. Films were deposited onto the substrates held at ~70°C.

[39,46,59–64] have also been used since they provide a clear measurement of the residual potential (electrostatic potential on the surface of a high-resistivity solid) due to trapped charges in the bulk.

As Tonchev et al. [15] assume, the improvement in the $\mu\tau$ product (Figure 6.13) with small additions of Sb implies that the deep trap population is reduced for both holes and electrons up to 1 at.% Sb. There is no substantial change in the microscopic mobility and capture coefficient. Although our structural studies by direct X-ray diffraction and indirect Raman scattering show that the structure of the films deposited onto substrates held at room temperatures and at ~70°C are nearly identical at the range of short- and medium-range order, the photoelectronic properties seem to be more sensitive to preparation conditions.

Figure 6.14 shows the temperature dependence of the drift mobility in various phases of selenium. It is apparent that the drift mobility at low temperatures in the liquid, monoclinic, and amorphous phases are thermally activated (the only exception is γ-Se). The hole drift mobility activation energy in amorphous and α-monoclinic Se is comparable. The basic interpretation of drift mobility-temperature data for

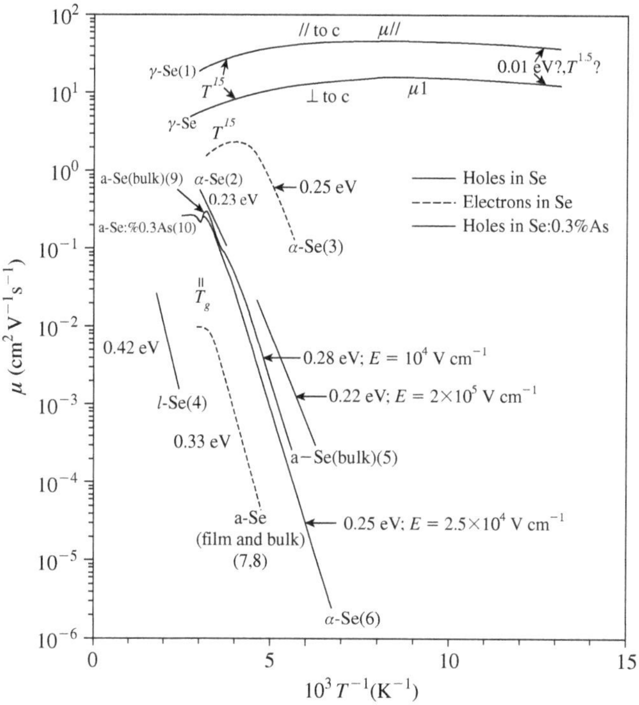

Figure 6.14 Hole and electron drift mobility in various phases of selenium: a-Se, amorphous Se, α-Se, α-monoclinic Se, γ-Se, trigonal Se, liquid Se (data collected by Kasap in [2]).

amorphous and α-monoclinic Se has been a shallow-trap-controlled transport mechanism. According to this mechanism, the hole drift mobility is expressed as

$$\mu_h = \mu_0(T)\frac{\tau_c}{\tau_c + \tau_i} = \mu_0(T)\left[1 + \frac{N_c}{N_v}\exp\left(\frac{E_t}{kT}\right)\right]^{-1},$$

where μ_0 is the microscopic mobility, τ_c and τ_i are the mean capture and mean release times, N_t is the shallow trap concentration, N_c is the density of states at the valence band mobility edge E_v, and E_t is the energy depth of the shallow traps. Although the exact nature of the basic microscopic conduction process has not been conclusively established, the present authors assume that it takes place by extended state transport (at least for the case of amorphous selenium).

The presently accepted model for the electronic density of states for a-Se shown in Figure 6.15 was developed by Abkowitz and coworkers [39,46,59–64]. The localized states, both shallow and deep, in the mobility gap are due to various structural defects that are thermodynamically stable at room temperature. Shallow traps of high concentration are located at the relatively discrete energies ~0.29 eV above E_v

and ~0.35 eV below E_c (Figure 6.15). These traps control hole and electron mobility. At present, shallow traps are known to be native defects. At the same time, their exact nature has not been conclusively determined. As proposed by Lucovsky and coworkers [65], they may be due to dihedral angle distortions in the random structure of a-Se in which the lone pair orbitals on adjacent Se atoms approach parallel alignment. The location of deep hole and electron traps can be determined from xerographic measurements (Figure 6.16).

It was well established by early TOF measurements that various impurities and alloying have drastic effects on the nature of charge transport in a-Se [9,66]. Appropriate chemical modification allows an optimum xerographic performance to be obtained from a-Se-based photoreceptor [2].

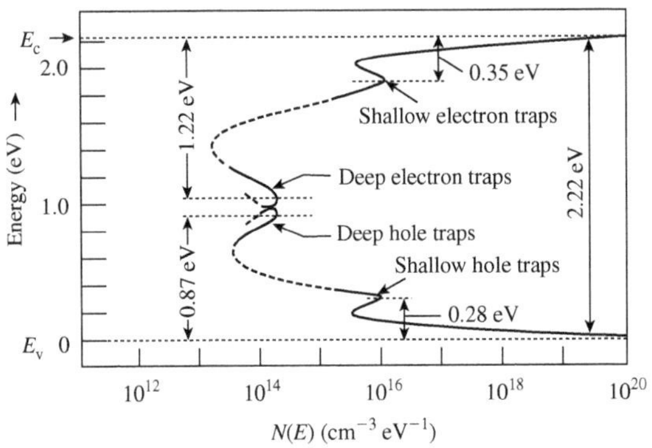

Figure 6.15 Density of states function $N(E)$ for a-Se [62].

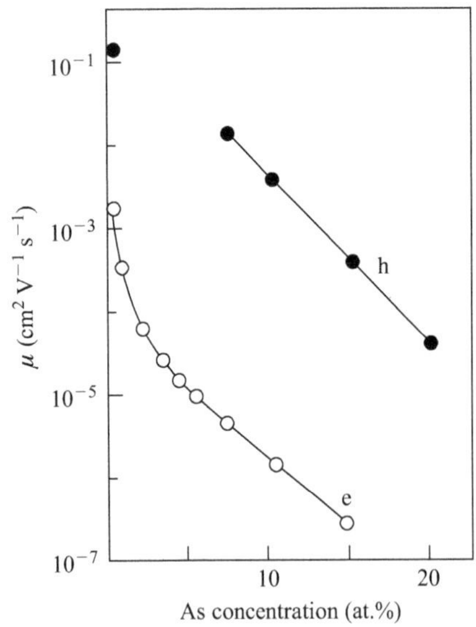

Figure 6.16 Effect of As alloying on drift mobility of holes (solid circles) and electrons (open circles) for amorphous selenium.

Owen and Spear [51] observed a similar effect of As addition/alloying on drift mobility of amorphous selenium. The same situation, we believe, is characteristic for Sb addition to a-Se [46].

Within the shallow trap-controlled transport model, the measured carrier range $\mu\tau$ is given by

$$\mu\tau = \mu_0\tau_0 = \mu_0/(C_t N_t),$$

where μ_0 is the microscopic mobility (in the extended states), τ_0 is the intrinsic deep trapping time in the absence of shallow traps, C_t is the capture coefficient, and N_t is the concentration of deep traps.

The experimentally observed peak of various physical parameters at about the same composition (1 at.% Sb) is remarkable.

6.5 PID Characteristics

In essence, the xerographic photosensitivity (S) of a photoreceptor material determines the rate of decay, dV/dt, of the electrostatic surface potential of the samples during PID. The xerographic photosensitivity definition adopted here is simply based on the amount of light energy required for the surface potential to decay to half of its original value ($V_0/2$) during PID (i.e., the fractional change in the surface potential per unit of light exposure).

Figure 6.17 shows a dark decay curve and a PID curve for the a-$Sb_{0.03}Se_{0.97}$ film. It can be seen that the sample exhibits relatively little dark decay. Nevertheless, in order to take into account the surface charge reduction during illumination and to evaluate S accurately, the contribution of dark discharge to the total change in the surface potential during PID was subtracted (V_0–V_d in Figure 6.17).

The xerographic spectral responses for both a-$Sb_{0.03}Se_{0.97}$ and pure Se are shown in Figure 6.18.

It is apparent that as Sb is added to a-Se, the photosensitivity at a particular wavelength increases. More precisely, this means that the xerographic photosensitivity for

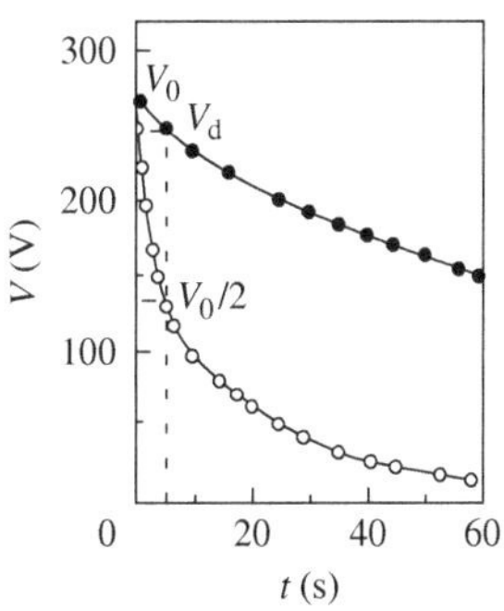

Figure 6.17 Dark decay curve (solid circles) and PID curve (open circles) for $Sb_{0.03}Se_{0.97}$.

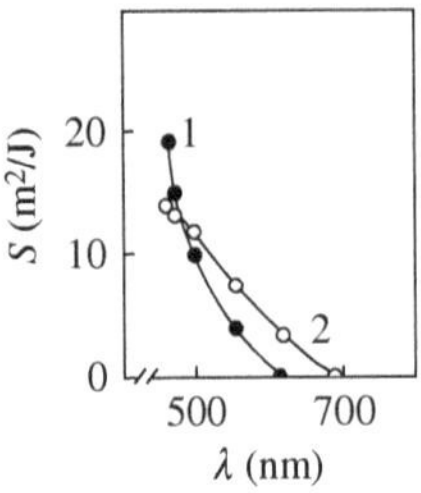

Figure 6.18 Xerographic photosensitivity (S) versus exposure wavelength (λ) for pure Se (1) and $Sb_{0.03}Se_{0.97}$ (2) alloy films.

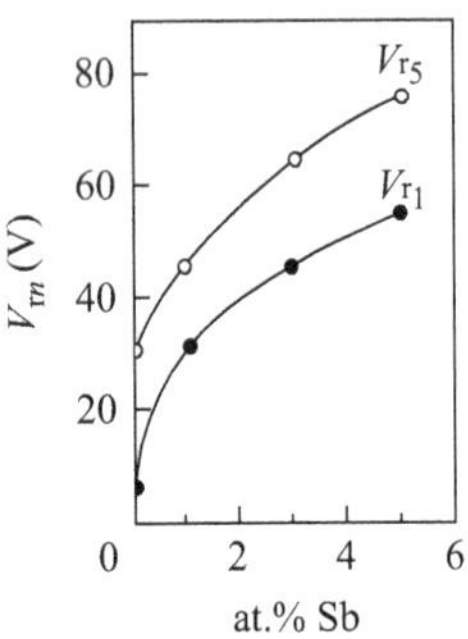

Figure 6.19 Effect of antimony on the residual potential (measured after first and fifth cycles) of Sb–Se alloy.

the Sb_xSe_{1-x} alloy is somewhat greater at longer wavelengths ($\lambda > 670$ nm) than for pure Se, and it is smaller at shorter wavelengths ($\lambda < 500$ nm). Note that other compositions of Sb–Se alloy showed similar trends. The result of increased xerographic photosensitivity of the Sb–Se alloys at longer wavelengths suggests that the addition of antimony to a-Se causes a reduction of the band-gap of the material. It should be noticed here that the xerographic photosensitivity depends not only on the absorption coefficient, α, but also on the quantum efficiency, η, for generating mobile charge carriers as well as the transport properties (the $\mu\tau$ product).

For the films under examination, the residual voltage V_r (a measurable surface potential at the end of the illumination) increases with Sb content (Figure 6.19).

The residual potential is due to trapped electrons in the bulk of the specimen. The simplest theoretical model, which is based on range limitation and weak trapping ($V_r \ll V_0$), relates V_r to $\mu\tau$ (the drift mobility, μ, and lifetime, τ, product) via the Warter equation [67]

$$V_r = L^2/(2\mu\tau),$$

where L is the sample thickness. For example, addition of 3 at.% Sb leads to a change in the first-cycle residual voltage from 4 to 44 V, which is equivalent to a change of the carrier range $\mu\tau$ from 10^{-7} to $10^{-6}\,cm^2V^{-1}$. Substituting $\mu^e \sim 7 \times 10^{-3}\,cm^2V^{-1}s^{-1}$ for pure Se and $\mu^e \sim 6 \times 10^{-4}\,cm^2V^{-1}s^{-1}$ for $Sb_{0.03}Se_{0.97}$ into the corresponding equation, we find carrier lifetimes $\tau \approx 2 \times 10^{-4}$ s and $\tau \approx 1.3 \times 10^{-3}$ s in a-Se and a-$Sb_{0.03}Se_{0.97}$. It is necessary to note here that in general,

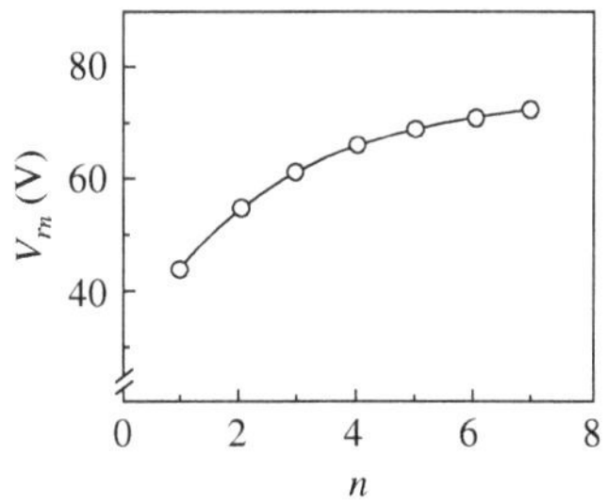

Figure 6.20 The buildup in the residual voltage with number of xerographic cycles in a-$Sb_{0.03}Se_{0.97}$ films.

bulk deep trapping lifetimes computed from the first-cycle residuals are in agreement with lifetimes measured in the TOF mode under range-limited conditions.

Figure 6.20 displays the buildup of the residual voltage V_{rn} on a-$Sb_{0.03}Se_{0.97}$ film with the number of xerographic cycles n. The rate of V_{rn} decreases with cycling. Then, for large n ($\geq$6 in our case), V_{rn} tends toward a saturation value V_{rs}.

As described in Chapter 5, the saturation residual potential provides an experimental measure of the integrated number of deep traps (trap-release rates are much slower than those from shallow traps which control drift mobility). V_{rs} is then simply given by

$$V_{rs} = eN_tL^2/(2\varepsilon),$$

where N_t is the deep-trap concentration, and ε is the dielectric constant.

Both the first residual and the cycled-up saturated residual potential are sensitive to alloying. For example, when pure amorphous Se films are alloyed with antimony, the buildup of the residual potential occurs more rapidly toward a much higher saturated residual potential. We obtain, for instance, $N_t \sim 2 \times 10^{14}\,cm^{-3}$ and $N_t \sim 10^{15}\,cm^{-3}$ for a-Se and $Sb_{0.03}Se_{0.97}$, respectively.

The photoelectric properties of a-Sb_xSe_{1-x} alloys can be at least qualitatively explained by using concepts based on charged structural defects, called valence alternation pairs (VAP) or intimate valence alternation pairs (IVAP). These correspond to some of the chalcogen atoms being under- and over-coordinated [68]. It seems reasonable that dark discharge and residual voltage buildup involve essentially the same species of localized centers. Further, it is also possible that we may be observing amphoteric behavior by IVAPs. For pure Se, an IVAP comprises over- and under-coordinated selenium atoms Se_3^+ and Se_1^- in close proximity. An IVAP center would be seen as a neutral trap by the carrier. The capture of an electron by Se_3^+ exposes the negative charge on Se_1^- and explains the residual voltage detected. At the same time, the emission of an electron from the Se_1^- uncovers the positive charge on Se_3^+, which causes the dark decay. Although the question remains whether the neutral centre is a D^0-type defect or whether it is an IVAP defect, the measured radius of 3 Å in Ref. [69] is representative of an IVAP capture radius.

There are a number of desirable electrical characteristics that a useful photoreceptor should exhibit. Our experimental results show that dark decay rate, residual voltage, and drift mobility are all sensitive to antimony. Carrier drift mobility decreases with antimony whereas dark decay rate and residual voltage increase. All these effects are undesirable in xerography. At the same time, the main advantage of a-Sb_xSe_{1-x} is that its spectral response can be readily shifted to longer wavelengths by increasing the Sb content, which allows for photoreceptor designs that can respond to a variety of illumination spectra. In addition, it may be possible to improve the charge transport parameters by halogen doping. Indeed, Cl is known to compensate for As-induced deep traps in a-Se (same group as Sb) [46]. By using a double-layer photoreceptor consisting of a thin a-Sb_xSe_{1-x} layer for photogeneration and a thick a-Se layer for charge transport, the xerographic parameters can be further improved.

Thus, the mobile carrier species controlling the xerographic depletion discharge in a-Sb–Se alloys are electrons. Thermal generation of free electrons in a-Sb_xSe_{1-x} is accompanied by the simultaneous formation of a deeply trapped positive space-charge. It was shown that antimony alloying progressively enhances the free electron thermal generation rate relative to the pure specimen.

As apparent from the large xerographic residual potentials for Sb_xSe_{1-x} alloys, the addition of Sb to a-Se seems to greatly increase the concentration of deep localized states within the mobility gap of the material.

The results indicate that in the long wavelength region (e.g., λ 600 nm), the photosensitivity for the a-Sb_xSe_{1-x} films is higher than for the pure selenium, probably due to a greater quantum efficiency.

6.6 Conclusion

It has been observed that the addition of antimony increases the DC conductivity, decreases the electron mobility, changes the xerographic parameters, and decreases the optical gap of the system. TOF experiments indicate that electron transport in a-Sb_xSe_{1-x} films is controlled by a set of shallow traps located at ~0.33 eV below E_c, the concentration of which increases with Sb addition. Xerographic dark discharge measurements showed that the decay of the surface potential is essentially due to bulk thermal generation of electrons and their subsequent sweep-out and depletion. Electron emission occurs from mid-gap localized states. When a-Se is alloyed with antimony, the dark discharge becomes more rapid due to an increase in the volume density of the mid-gap electron emission centers with antimony concentration. Cycled-up xerographic residual voltage measurements indicate that the saturated residual voltage increases with Sb addition and is caused by an increase in the concentration of electron deep traps ($N_t \sim 10^{15}\,\mathrm{cm}^{-3}$). The xerographic photosensitivity for Sb_xSe_{1-x} alloys is greater at longer wavelengths, $\lambda > 670$ nm, than for pure Se. The compositional dependence of the electronic properties is related to incorporation of Sb atoms in a glassy selenium matrix. Finally, the most essential result is the extremely high X-ray sensitivity of Sb-containing amorphous selenides.

6.7 References

1. A. Madan and M.P. Shaw, The Physics and Applications of Amorphous Semiconductors (Academic Press, Boston, MA, 1988).
2. S.O. Kasap, Handbook of Imaging Materials (eds. A.S. Diamond and D.S. Weiss) (Marcel Dekker Inc., New York, NY, 2nd ed., 2002).
3. K. Tanaka, Curr. Opin. Solid State Mater. Sci., 1 (1996) 567.
4. K. Tanaka, Encyclopedia of Materials: Science and Technology (Elsevier Science Ltd., New York, Basel, 2001) 1123–1131.
5. J. Rowlands and S. Kasap, Phys. Today 11 (1997) 24.
6. V. Damodara Das, K.S. Raju, S. Aruna, J. Appl. Phys. 78 (1995) 1751.
7. M. Fadel, M.M. El-Samanoundy, K.A. Sharaf, J. Mater. Sci. 30 (1995) 2377.
8. S.O. Kasap, J. Rowlands, B. Fogal, M. Zahangir Kabir, G. Belev, N. Sidhu, B. Polischuk, R.E. Johanson, J. Non-Cryst. Solids 299–302 (2002) 988.
9. J. Schottmiller, M. Tabak, G. Lucovsky, A. I. Ward, J. Non-Cryst. Solids 4 (1970) 80.
10. M. Abkowitz and J.M. Markovich, Solid State Commun. 44 (1982) 1431.
11. N.F. Mott and A.E. Davis, Electronic Processes in Non-Crystalline Materials (Clarendon Press, Oxford, 1979).
12. C. Wood, R. Mueller, L.R. Gilbert, J. Non-Cryst. Solids 12 (1973) 295.
13. M.H. El-Zaidia, A. El-Shafi, A.A. Ammar, M. Abo-Ghozala, Thermochim. Acta 116 (1987) 35.
14. D. Tonchev, S.O. Kasap, J. Non-Cryst. Solids 248 (1999) 28.
15. D. Tonchev, B. Fogal, G. Belev, R.E. Johanson, S.O. Kasap, J. Non-Cryst. Solids 299–302 (2002) 998.
16. V.I. Mikla, I.P. Mikhalko, Y.Y. Nagy, J. Phys.: Condens. Matter 6 (1994) 8269.
17. V.I. Mikla, Phys. Status Solidi B 182 (1994) 325.
18. V.I. Mikla, Y.Y. Nagy, V.V. Mikla, A.V. Mateleshko, Mater. Sci. Eng. B 64 (1999) 1.
19. V.I. Mikla, A.V. Mateleshko, V.V. Mikla, Y.Y. Nagy, J. Non-Cryst. Solids 246 (1999) 46.
20. F.V. Pirogov, J. Non-Cryst. Solids 114 (1989) 76.
21. W.E. Spear, J. Non-Cryst. Solids 1 (1969) 197.
22. G. Pfister and H. Scher, Adv. Phys. 27 (1978) 747.
23. S. Mahadevan, A. Giridhar, A.K. Singh, J. Non-Cryst. Solids 88 (1986) 11.
24. M. Reading, Trends Polym. Sci. 1 (1993) 248.
25. K.J. Jones, I. Kinshott, M. Reading, A.A. Lacey, C. Nicopoulos, M.M. Polosk, Thermochim. Acta 304 & 305 (1997) 187.
26. V.I. Mikla, I.P. Mikhalko, Y.Y. Nagy, A.V. Mateleshko, V.V. Mikla, J. Mater. Sci. 35 (2000) 4907.
27. Z.U. Borisova, Glassy Semiconductors (Plenum Press, New York, NY, 1981).
28. J.C. Phillips, J. Non-Cryst. Solids 44 (1981) 17.
29. M.F. Thorpe, J. Non-Cryst. Solids 57 (1983) 355.
30. M.F. Thorpe, J. Non-Cryst. Solids 76 (1985) 109.
31. J.C. Phillips, M.F. Thorpe, Solid State Commun. 53 (1985) 699.
32. V.I. Mikla, J. Phys.: Condens. Matter 9 (1997) 9209.
33. A.R. Melnyk, J. Non-Cryst. Solids 35–36 (1980) 837.
34. M. Abkowitz, G.M.T. Foley, J.M. Markovics, A.C. Palumbo, Appl. Phys. Lett. 46 (1985) 393.
35. M. Baxendale, C. Juhasz, SPIE Proc. (1990) 1253.
36. S.O. Kasap, M. Baxendale, C. Juhasz, IEEE Trans. Indust. Appl. 27 (1991) 620.

37. S.O. Kasap, J. Electrostat. 22 (1989) 69.
38. M. Abkowitz, F. Jansen, A.R. Melnyk, Phil. Mag. B 51 (1985) 405.
39. M. Abkowitz, J. Non-Cryst. Solids 66 (1984) 315.
40. M. Abkowitz, R.C. Enck, Phys. Rev B 25 (1982) 2567.
41. M. Abkowitz, R.C. Enck, Phys. Rev B 27 (1983) 7402.
42. S.M. Vaezi-Nejad, PhD Thesis, University of London, UK (1984).
43. I.P. Batra, K.K. Kanazawa, H. Seki, J. Appl. Phys. 41 (1970) 3416.
44. S.E. Berger, R.C. Enck, Proc. of the Symposium of Industrial uses of Selenium and Tellurium (Toronto, Canada, 1980) 166.
45. S.M. Vaezi-Nejad, Int. J. Electronics 62 (1987) 361.
46. M. Abkowitz Physics of Disordered Materials (eds. D. Adler, H. Fritzsche, S.R. Ovshinsky), (Plenum Press, New York, NY, 1984) 483.
47. J.M. Marshall, Rep. Progr. Phys. 46 (1983) 1235.
48. J.R. Haynes and W. Shockley, Phys. Rev. 75 (1949) 691.
49. W.E. Spear, Proc. Phys. Soc. (London) B76 (1960) 826.
50. W.E. Spear, W.E. Steemers, H. Mannsperger, Philos. Mag. B 48 (1983) L49.
51. A.E. Owen, W.E. Spear, Phys. Chem. Glasses 17 (1976) 174.
52. O.V. Luksha, V.I. Mikla, V.P. Ivanitsky, J. Non-Cryst. Solids 67 (1989) 1223.
53. J.M. Marshall, F.D. Fisher, A.E. Owen, Phys. Status Solidi A25 (1974) 419.
54. R.A. Street, Appl. Phys. Lett. 41 (1982) 1060.
55. R.A. Street, J. Zesch, M.I. Thompson, Appl. Phys. Lett. 43 (1983) 1425.
56. R.A. Street, J. Zesch, M.I. Thompson, Appl. Phys. Lett. 43 (1983) 672.
57. R.A. Street, C.C. Tsai, M. Stutzman, J. Kakalios, Philos. Mag. B 56 (1987) 389.
58. Z. Hecht, Z. Phys. 77 (1932) 235.
59. M. Abkowitz, Polym. Eng. Sci. 24 (1984) 1149.
60. M. Abkowitz, J. Non-Cryst. Solids 77–78 (1985) 191.
61. M. Abkowitz, J. Non-Cryst. Solids 97–98 (1987) 1163.
62. M. Abkowitz, Philos. Mag. Lett. 58 (1988) 53.
63. M. Abkowitz, Ann. N. Y. Acad. Sci. 371 (1981) 171.
64. M. Abkowitz, I.M. Markovics, Philos. Mag. B 49 (1984) L31–L36.
65. C.K. Wong, G. Lucovsky, J. Bernholc, J. Non-Cryst. Solids 97–98 (1985) 1171.
66. B.T. Kolomiets, E.A. Lebedev, Sov. Phys. Solid State 8 (1966) 905.
67. P.J. Warter, Appl. Optic. Suppl. 3 (1969) 65.
68. M. Kastner, D. Adler, H. Fritzsche, Phys. Rev. Lett. 37 (1976) 1504.
69. B. Polischuk, S.O. Kasap, V. Aiyah, M. Abkowitz, Can. J. Phys. 69 (1991) 364.

7 High-Definition TV Pickup Tubes

Development of high-definition TV (HDTV) pickup devices is reviewed. The amorphous chalcogenide Saticon and the more recent HARP (high-gain avalanche rushing amorphous photoconductor) photoconductive surface are suitable for use in HDTV since both have an inherently high resolving power and low lag characteristics. Experimental pickup tubes have been considered. Special attention has been paid to the HARP pickup tube, as it has shown that it can satisfy most requirements for the image device of the HDTV system, such as resolving power, lag characteristics, and signal uniformity.

Extensive efforts have been devoted to the development of an HDTV system, which is expected to offer high-quality pictures not only for home TV use but also for electronic cinema production, electronic printing, medical diagnostics, etc. [1–3].

In order to realize this system, highly technological innovations were required mainly in the fields of color cameras, video recorders, and color displays. The development of a high-definition color camera was an especially challenging subject since it included the realization of a new image pickup device. Every conventional image pickup device for broadcasting use, either pickup tube or solid-state imager, needed more than 10 years to reach its present stage of development. This is because an image device for broadcast use should fulfill many of the characteristic conditions that are sometimes in conflict with each other.

A high-definition image pickup device should have, first of all, high resolving power for up to 2,000 TV lines. This value is roughly equivalent to realizing a 4 Mbit random access memory in solid-state devices. If we drive this device with 60 fields per second, the clock frequency will be 240 MHz. It can be understood that these requirements are too severe to be accomplished by the present stage of silicon device technology. High-resolution pickup tubes have past records in resolving power, but their lag characteristics have been too poor for broadcasting use.

The vidicon is a storage-type camera tube in which a charge-density pattern is formed by the imaged scene radiation on a photoconductive surface, which is then scanned by a beam of low-velocity electrons. The fluctuating voltage coupled out to a video amplifier can be used to reproduce the scene being imaged. The electrical charge produced by an image will remain in the faceplate until it is scanned or until the charge dissipates. Pyroelectric photocathodes can be used to produce a vidicon sensitive to a broad portion of the infrared spectrum. Investigation of an electron-optical system in vidicon-type pickup tube, however, has shown that an almost satisfactory image device for HDTV can be realized by using a Se–As–Te Saticon target.

Amorphous Chalcogenides. DOI: 10.1016/B978-0-12-388429-9.00007-8

Kenkichi Tanioka and coworkers from NHK (Hitachi) have developed the world's first imaging device that achieves ultrahigh sensitivity and high picture quality by using a continuous and stable avalanche multiplication phenomenon in an amorphous selenium photoconductive target. The amorphous photoconductive target operating in the avalanche mode is called HARP. Recently, HARP film used for the red channel in a color camera was doped with Te to improve its red-light sensitivity [4]. A heat treatment to suppress the defect occurrence was found.

In addition, Ohkawa et al. [5] developed a 35-μm- thick HARP film, which was thicker than the previous 25-μm- thick film.

7.1 Saticon

Successful application of amorphous Se–As–Te alloys is in Hitachi's Saticon [2]. It is a commercially available TV pickup tube (a vidicon-type image device). The structure of a typical a-Se–As–Te Saticon, which utilizes the high panchromatic photosensitivity of the Se–Te–As alloy and the relatively fast hole drift mobility of a-Se, is shown in Figure 7.1. It is basically a heterostructural thin-film photodiode formed between *n*-type tin oxide and *p*-type selenium-rich amorphous chalcogenide. In order to block hole injection from the SnO_2 into the amorphous chalcogenide, a very thin layer of GeO_2 is inserted between them. A porous Sb_2S_3 layer prevents secondary electron emission into vacuum as well as electron injection into the amorphous chalcogenide. The physical behavior of this blocking diode structure may be summarized as follows: photogenerated carriers in the Se–As–Te layer freely flowing out, whereas carrier injection into the layer is blocked. As most of the applied voltage is sustained by this highly resistive Se–As–Te layer, a fast response due to a primary drift photocurrent is expected.

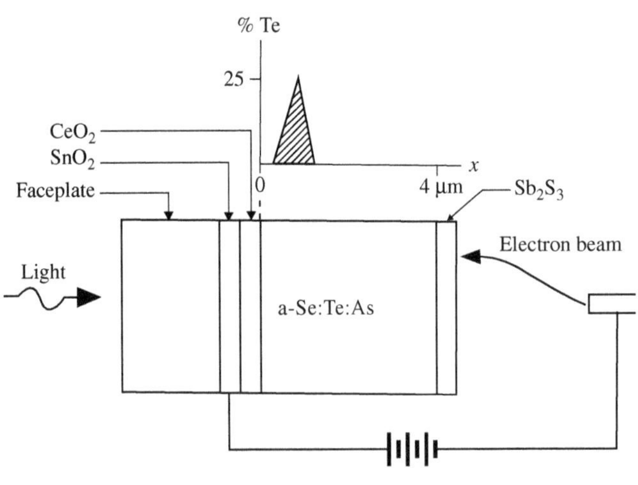

Figure 7.1 Schematic target structure of the Saticon pickup tube.

In order to enhance the red-light sensitivity, a Te-rich region in the Se–As–Te layer is formed. Most of the photocarriers are generated in this very thin region of less than 100 nm.

The target structure of Saticon is especially suited for the high-definition pickup device for the following reasons:

1. The amorphous Se–As–Te photoconductor has essentially high resolving power, because it has neither grain boundaries nor any microstructures, which deteriorate the spatial uniformity. The intrinsic resolution of the layer will be less than 5 nm.
2. Most of the incident photons are absorbed within the very thin Te-rich region and generate photocarriers there. These carriers are extracted out of the region and drifts by the applied electric field as high as 10^5 V/cm. The lateral diffusion of photocarriers during their transit to the surface of the layer, therefore, will be very small compared with the thickness of the layer.
3. The optical absorption coefficient of the Se–As–Te layer is relatively high, and its optical reflectance is low compared with other practical photoconductors like PbO. This is favorable for suppressing the undesirable flare effect, which comes from optical multireflection within the photoconductive layer or between the layer and the optical system.

The realization of a high-definition pickup tube, therefore, can be achieved by developing a suitable electron-optical system, which does not deteriorate the inherently high resolution and slow response speed of the Se–As–Te target.

However, sensitivity of Saticon imaging devices was still rather poor, and as current affairs programs began to attach increasing importance to camera sensitivity, even for standard resolution TV broadcasts, it became even more important to develop a TV camera with high sensitivity and high picture quality that could produce clear images from poorly lit subjects.

Consequently, from about 1980, Tanioka's group began a fresh study with the aim of realizing a high sensitivity and high picture quality imaging device suitable for use in HDTV applications. This study focused on using an amplification effect on an amorphous selenium target (photoelectric conversion film) to obtain a high level of sensitivity.

7.2 Target Structure and Current–Voltage Characteristics

In 1985, Tanioka found that when the a-Se target operates in a strong electric field of about 10^8 V/m, continuous and stable avalanche multiplication takes place, allowing high sensitivity to be obtained with little picture degradation. Based on this discovery, Tanioka's group went on to develop a new kind of pickup tube called HARP, which has been studied to this day to further increase sensitivity and achieve a wider range of applications [4,5].

HARP pickup tubes, which have achieved sensitivities roughly 100 times greater than CCDs, are used not only in HDTV handheld cameras but also in the production of nighttime news flashes and special programs, such as imaging the aurora.

A schematic diagram of the experimental tube target is shown in Figure 7.2.

The photosensitive layer is of evaporated amorphous selenium. The thickness of the target layer is 2 μm. A thin layer of antimony trisulfide is deposited on the scanning side of the photoconductor to prevent electron injection from the scanning beam and to reduce the emission of the secondary electrons. Between the selenium layer and the signal electrode, a thin layer of GeO_2 (about 20 nm) is interposed to make the hole-blocking contact stable. These layers are deposited using an ordinary vacuum-evaporation method. This target can be operated as a blocking target.

Figure 7.3 shows the signal current and the dark current versus the target voltage (V_t).

Blue light was incident on the target. The signal current rapidly increases and eventually saturates with increasing target voltage. If the target voltage is increased

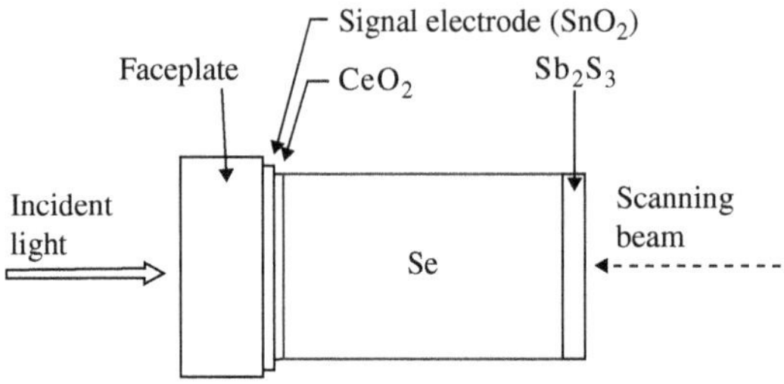

Figure 7.2 Schematic diagram of experimental tube target [6].

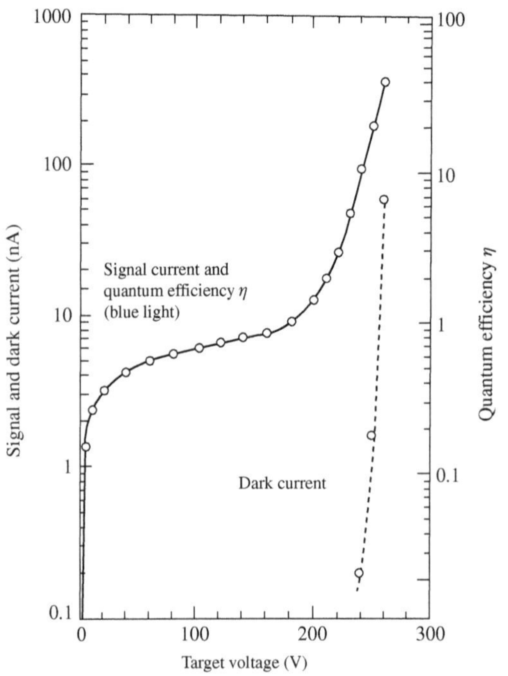

Figure 7.3 Signal current and dark current versus target voltage.

further, the signal current again increases rapidly. The quantum efficiency $\eta = 40$ at $V_t = 260\,V$ indicates a very high sensitivity.

In the region in which the sensitivity again increases rapidly, the dark current also increases. However, at a target voltage of 240 V ($\eta \approx 10$), the dark current is as little as 0.2 nA. These characteristics imply that some multiplicative phenomenon occurs in the amorphous selenium layer.

7.3 Properties of the Multiplicative Phenomenon

Figure 7.4 shows the effective storage capacitance versus target voltage for an experimental target.

The capacitance was calculated from the signal current, which was generated by changing the cathode potential with pulses of 1 V for each target voltage.

The capacitance does not depend on the target voltage. If electron injection occurred from the scanning beam to the photoconductor, the capacitance would increase with increasing target voltage. This implies that the multiplicative phenomenon is not caused by electron injection.

Figure 7.5 shows the signal current for photogenerated holes and electrons versus the target voltage in an experimental tube.

When the target is illuminated through the faceplate by blue light, holes are the dominant carriers in the photogenerated signal current. On the other hand, when the target is illuminated from the scanning beam side, electrons are the dominant carriers. Because the signal electrode is biased positively with respect to the scanning electron beam, the carriers are generated in the surface region (this is thin compared with the a-Se layer). The hole current and the electron current show almost the same characteristics up to a target voltage of 160 V. However, at target voltages of more than 160 V, the hole current is much larger than the electron current.

Figure 7.6 shows the current–field characteristics of experimental tubes with a-Se layers of different thicknesses.

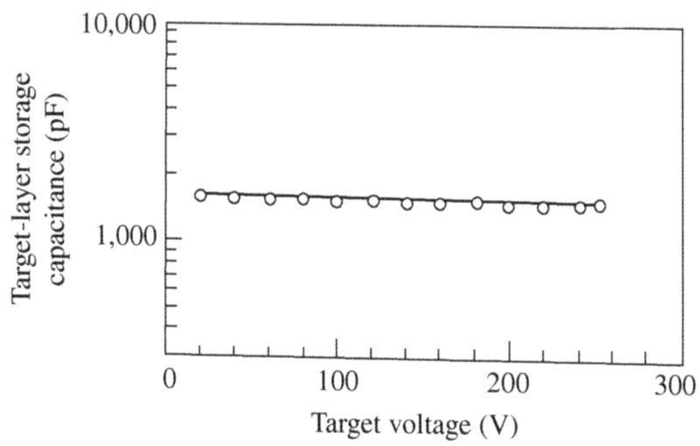

Figure 7.4 Effective storage capacitance versus target voltage for an experimental target.

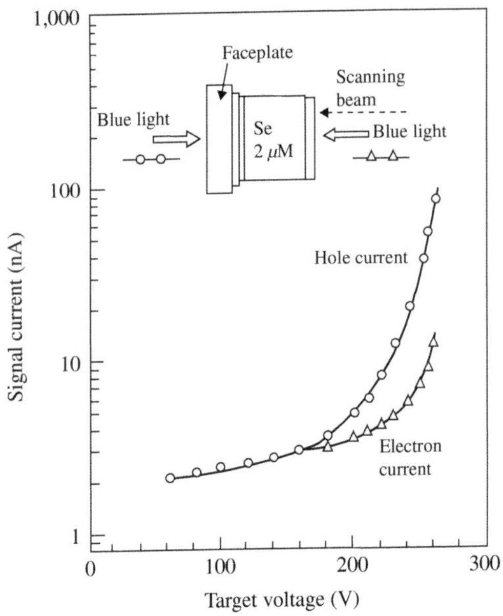

Figure 7.5 Signal current for photogenerated holes and electrons versus the target voltage.

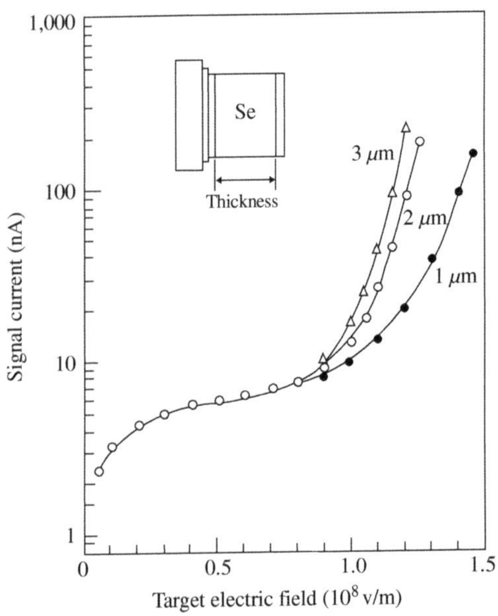

Figure 7.6 Current–field characteristics of experimental tubes with a-Se layers of different thicknesses.

7.4 New Super-HARP Pickup Tube

Tanioka et al. [7,8] have recently developed a greatly improved version of the HARP tube with a selenium layer 25 μm thick because sensitivity as a function of the target's electric field increases with the target thickness.

An operational principle of the HARP tube is shown in Figure 7.7. The light energy absorbed in the selenium target generates an electron–hole pair. The carriers are accelerated by a large electric field, 10^8 V/m, and then the hole, which has increased kinetic energy, generates a new electron–hole pair by means of impact ionization. This phenomenon occurs again and again throughout the target. The additional noise produced by avalanche multiplication is negligible so that the tube has high sensitivity.

Except for the thickness and the amount of Te in the selenium layer, the target structure inside the beam scanning area of the New Super-HARP is almost the same as in the conventional HARP target (Figure 7.8).

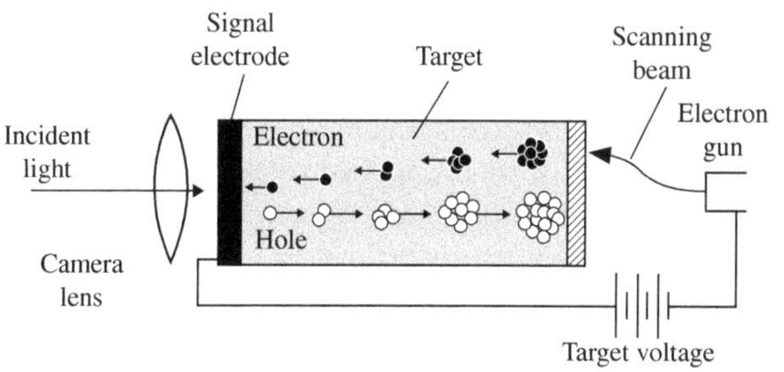

Figure 7.7 Operational representation of the HARP tube.

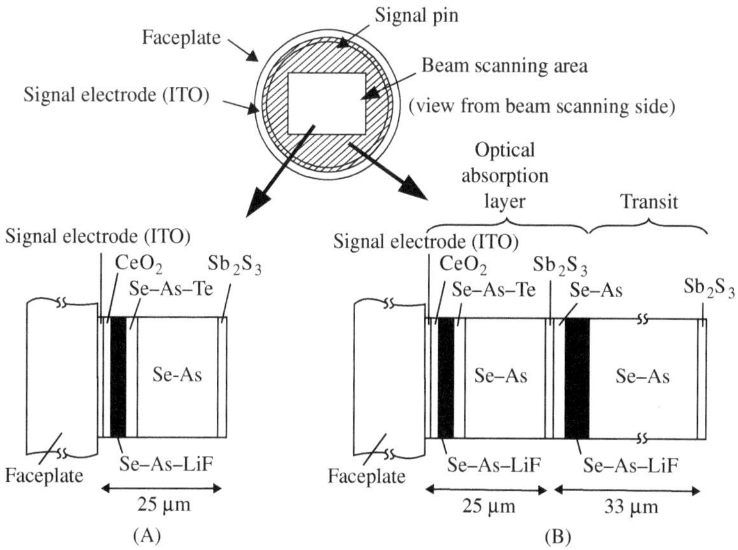

Figure 7.8 Structure of the New Super-HARP target (A) inside the beam scanning area and (B) outside the beam scanning area.

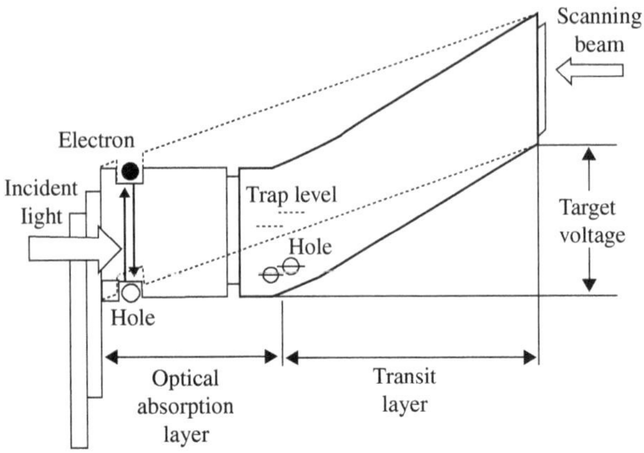

Figure 7.9 Band model of the insensitive target.

However, a newly designed structure outside the scanning area of the target stabilizes the beam inside the scanning area even when the applied voltage is very high.

Basically, the insensitive target was fabricated by adding a selenium layer with arsenic to the sensitive target. Lithium fluoride was doped near the boundary surface of the extra layer. A band model of this insensitive target is shown in Figure 7.9.

Layers of GeO_2 and Sb_2S_3 act as hole- and electron-blocking contacts. Electrons injected by the scanning electron beam become trapped in the Sb_2S_3 layer. These trapped carriers formed a negative space charge in the layer. Photogenerated holes in a-Se, from exposure to the image, transit across toward Sb_2S_3 to recombine with the electrons trapped in Sb_2S_3. The photogeneration is contained in a region of high Te concentration. The crystallization of a-Se is inhibited by the addition of ~1% As to a-Se. Saticon uses an amorphous material with distinct advantages (it is grainless as well as uniform material), and it exhibits high resolution.

Figure 7.10 shows the signal current and dark current versus target voltage in the New Super-HARP pickup tube. The incident light was green. The signal current rapidly increases at target voltages of more than 1,800V. This phenomenon resulted from avalanche multiplication in the selenium layer of the target. The figure shows that an avalanche multiplication factor of several hundred can be obtained at a target voltage of 2,500V.

Figure 7.11 shows the spectral response characteristics of the New Super-HARP tube. The quantum efficiency of the New Super-HARP pickup tube for the green incident light (540nm) was found to be double that of the selenium layer without tellurium.

Figure 7.12 shows a monitor picture produced by the three-tube color camera. To illustrate the big difference in sensitivity between the New Super-HARP camera and a CCD camera, Figure 7.12B shows a picture taken under the same conditions with a three-CCD color camera.

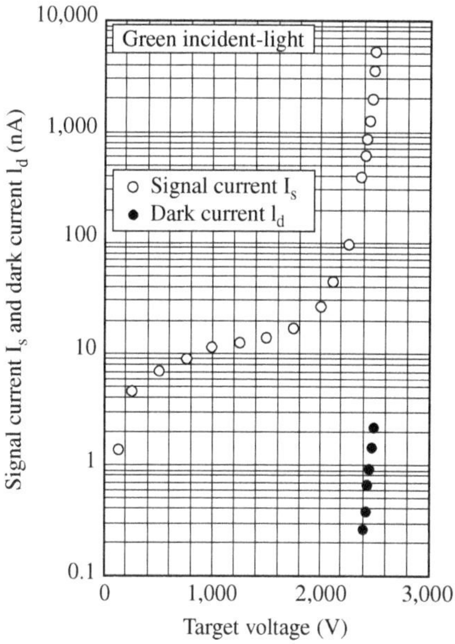

Figure 7.10 Signal current and dark current versus target voltage in the New Super-HARP pickup tube.

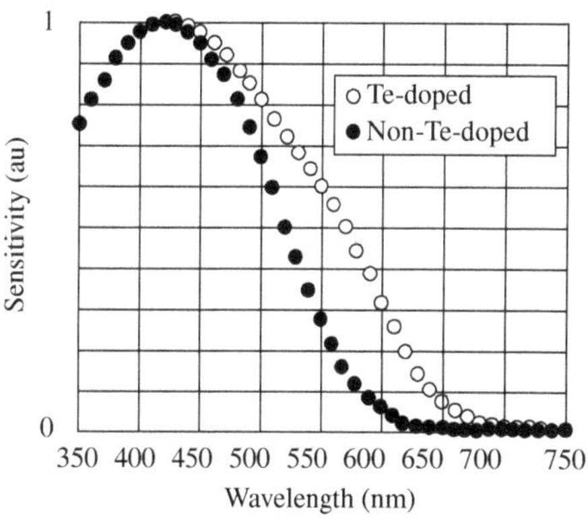

Figure 7.11 Spectral response characteristics of the New Super-HARP pickup tube.

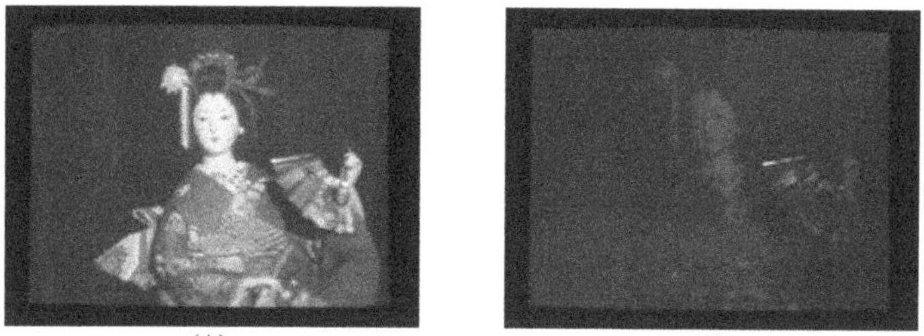

(A) (B)

Figure 7.12 Monitor pictures produced by color cameras with New Super-HARP tubes (A) and CCDs (B). Illumination is 0.3 lx.

7.5 Conclusion

We have discussed Hitachi's previous and present successful research into ultrahigh sensitivity HARP pickup tubes and the application of these tubes.

In general, these ultrahigh sensitivity imaging devices have so far been regarded as special-purpose devices in a separate class from ordinary high picture quality imaging devices such as CCDs. These ultrahigh sensitivity devices are able to take clear pictures with less noise than any other existing device under all lighting conditions.

7.6 References

1. K. Tanioka, J. Mater. Sci. Mater. Electr. 18 (2007) S321.
2. N. Goto, Y. Isozaki, K. Shidara, E. Maruyama, T. Hirai, T. Fujita, IEEE Trans. Electron. Dev. ED-21 (1974) 662.
3. E. Maruyama, Jpn. J. Appl. Phys. 21 (1982) 213.
4. Y. Ohkawa, K. Miyakawa, T. Matsubara, K. Kikuchi, S. Suzuki, K. Tanioka, M. Kubota, N. Egami, A. Koboyashi, IEICE Electron. Express 6 (2009) 1118.
5. Y. Ohkawa, K. Miyakawa, T. Matsubara, K. Kikuchi, S. Suzuki, M. Kubota, N. Egami, A. Koboyashi, IEICE Trans. Electron. E 92-C (2009) 894.
6. K. Tanioka, J. Yamazaki, K. Shidara, K. Taketoshi, T. Kawamura, T. Horai, Y. Takasaki, Adv. Electron. Electron Phys. 74 (1988) 379.
7. K. Tanioka, T. Matsubara, Y. Ohkawa, K. Miyakawa, S. Suzuki, T. Takahata, N. Egami, K. Ogusu, A. Kobayashi, T. Hirai, T. Kawai, M. Hombo, and T. Yoshida, IEICE Trans. Electron. E86-C (2003) 1790.
8. K. Tanioka, Photon. Broadcasting Technol. 1 (2005) 1–25.

8 X-Ray Photoconductors for Direct Conversion of Digital Flat-Panel X-Ray Image Detectors

8.1 Principles of the Direct-Conversion Digital X-Ray Image Detector

If a radiologist were to speculate about the ideal kind of X-ray imaging system, what might come to mind is a digital flat-panel system that can perform all clinically important radiographic techniques at reduced dose. Immediately after the patient's X-ray exposure, it would provide a high-quality radiograph on a video monitor and would also be usable for real-time imaging (fluoroscopy). The physical form of the system would be similar to a film/screen cassette so that it would easily fit into current medical X-ray systems, as illustrated in Figure 8.1.

Presently, two essential methods have been adopted for digital radiography. Both are based on the use of phosphors. This means that they involve indirect conversion from X-ray photons to a detectable charge signal. The first is the digitalization of a signal from a video camera optically coupled to an X-ray image intensifier (cesium iodide phosphor). The second is the photostimulable phosphor system, commonly called the computed radiography (CR) system, which captures the latent image within a storage phosphor layer, which is subsequently read by a laser scanner. The

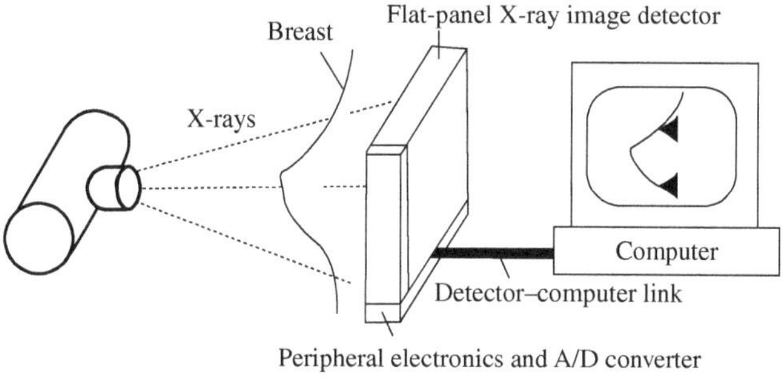

Figure 8.1 Schematic illustration of a flat-panel X-ray image detector for digital mammography. Connection from the detector to a local or distant computer is a convenient communication link [1].

Amorphous Chalcogenides. DOI: 10.1016/B978-0-12-388429-9.00008-X

intensifier system permits instantaneous readout. It is very bulky while the CR system, like film/screen, requires carrying the cassette from a loading/unloading station to the patient examination room and back. Neither of these indirect-conversion phosphor-based systems has adequate image quality for all applications. The need remains for a digital radiography system that reads out images electronically and directly with better image quality.

Recent research carried out concurrently at Sunnybrook Health Science Centre [1–4], DuPont [5,6], and Philips [7] has shown clearly that one of the most promising flat-panel digital radiographic systems is based on using a large-area thin-film transistor (TFT), active-matrix array (AMA) with an electrode X-ray photoconductor, as illustrated in Figures 8.2 and 8.3. An important first step in flat-panel thin-film technology was the development of flat-panel TFT displays, which matured as the fabrication and doping of large-area hydrogenated amorphous silicon (a-Si:H) films became technologically possible in the early 1990s [8]. Once large-area flat panels with small pixel sizes became available at the component level, it was only a matter of time before an X-ray photoconductor such as amorphous selenium would be used to directly convert X-ray images to a charge distribution stored on the pixels of a flat panel. Amorphous selenium photoconductor–based flat-panel X-ray detectors are now at a stage where they can provide excellent images, and they are being employed in various X-ray imaging applications—medical and nonmedical. The combination of an active array matrix and an X-ray photoconductor then constitutes a direct-conversion X-ray image detector, a term coined by DuPont (the terms *X-ray sensor* and *detector* have been used interchangeably in the literature). Direct conversion here refers to the fact that the X-ray photons are directly converted to charges and detected vis-à-vis an intermediate conversion, via a phosphor, to photons (light). They are then converted from photons to charges. The charge distribution residing on the panel's pixels are simply read out by scanning the arrays row-by-row using the

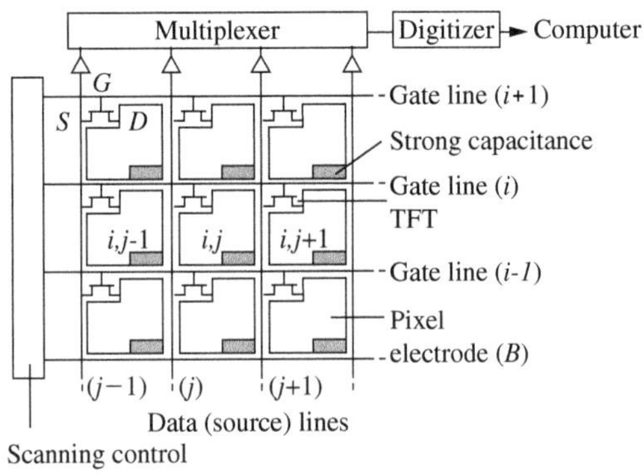

Figure 8.2 TFT AMA for use in X-ray image detecting with self-scanned electronic readout. The charge distribution residing on the panel's pixels are simply read out by scanning the arrays row-by-row using the peripheral electronics and multiplexing the parallel columns to a serial digital signal [3].

peripheral electronics and multiplexing the parallel columns to a serial digital signal, as illustrated in Figure 8.2.

The signal is then transmitted to a computer system. The system is simple, inherently digital, and has so many advantages that it has now become a major contending choice in digital radiography [3].

An AMA (Figure 8.2) consists of millions of individual pixel electrodes connected by TFTs (one for each pixel) to electrodes passing over the whole array to subsidiary electronics on the periphery. The TFTs act as switches to control the clocking out of image charge a line at a time. Very large-area (e.g., 30 cm × 30 cm) AMAs are now becoming available, and even larger ones will be possible in the future. The AMA consists of M × N (e.g., 1,280 × 1,536) TFT-based pixels, and each pixel (i, j) carries a charge collection electrode B connected to a signal storage capacitor C_{ij} whose charge can be read out by properly addressing the TFT (i, j) via the gate (i) and source (j) lines. External readout electronics and software, by proper self-scanning, converts the charges read on each C_{ij} to a digital image. Self-scanning here refers to the fact that no external means, such as a scanning laser beam as in some other X-ray imaging systems, is used to scan the pixels and extract the information [1,3]. The scanning operation is part of the flat-panel detector electronics and its software.

The a-Se layer is coated onto the AMA to serve as an X-ray photoconductor, as shown in Figure 8.3. An electrode (labeled A) is subsequently deposited on the a-Se layer to enable the application of a biasing potential and hence an electric field E in the a-Se layer. The EHPs that are generated in the photoconductor by the absorption of an X-ray photon travel along the field lines. Electrons are collected by the positive bias electrode (A) and holes accumulate on the storage capacitor C_{ij} and thereby provide a charge signal ΔQ_{ij} that can be read during self-scanning.

Each pixel electrode carries an amount of charge ΔQ_{ij} that is proportional to the amount of incident X-ray radiation by virtue of the X-ray photoconductivity of

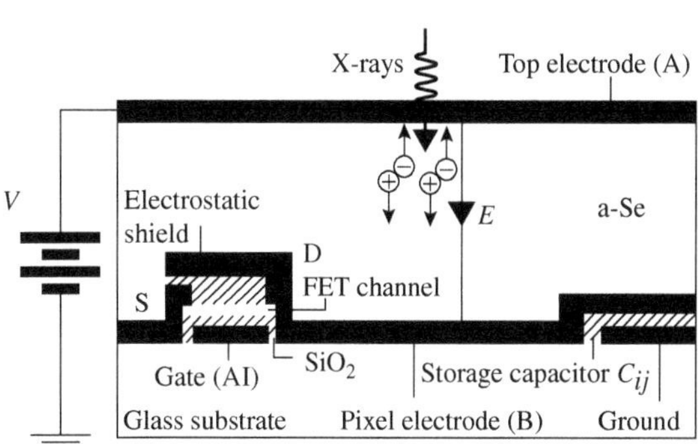

Figure 8.3 Cross section of a single pixel (i, j) with a TFT showing the accumulation of X-ray-generated charge on the pixel electrode and hence the storage capacitance C_{ij}. The top electrode (A) on the photoconductor is a vacuum-coated metal (Al). The bottom electrode (B) is the pixel electrode that is one of the plates of the storage capacitance (C_{ij}) [3].

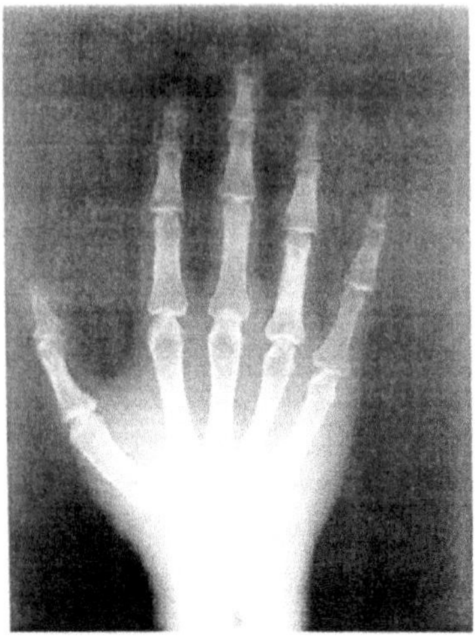

Figure 8.4 An X-ray image of a phantom hand obtained from a direct-conversion flat-panel image detector that uses a stabilized a-Se photoconductor [3].

the photoconductor over that pixel. All TFTs in a row have their gates connected, whereas all TFTs in a column have their sources connected. When gate line i is activated, all TFTs in that row are turned on, and N data lines from $j = 1$ to N then read the charges on the pixel electrodes in row i. The parallel data is multiplexed into serial data, digitized, then fed into a computer for imaging. The scanning control then activates the next row, $i + 1$, and all the pixel charges in this row are then read and multiplexed, and so on until the whole matrix has been read from the first to the last row (row M). It is apparent that the charge distribution residing on the panel's pixels is simply read out by self-scanning the arrays row-by-row and multiplexing the parallel columns to a serial digital signal. This signal is then transmitted to a computer system.

Figure 8.4 shows a typical X-ray image of a phantom hand obtained by a flat-panel X-ray image detector using an a-Se photoconductor as developed by Sterling Diagnostic Imaging (formerly DuPont). The resolution is determined by the pixel size, which in present experimental image detectors is typically $\sim 150\,\mu m$ but is expected to be as small as $50\,\mu m$ in future detectors to achieve the resolution necessary for mammography. It is interesting that the a-Se-based flat-panel X-ray sensor technology has been made possible by the use of two key elemental amorphous semiconductors: a-Si:H and a-Se. Although their properties are different, both can be readily prepared in large areas, which is essential for an X-ray image detector. It would be impractically difficult and expensive to develop a large-area detector using single crystal technology [3].

Any flat-panel X-ray image detector design must first consider the required specifications based on the clinical need of the particular imaging modality, for example mammography, chest radiology, and fluoroscopy.

Table 8.1 summarizes the specifications for flat-panel detectors for chest radiology, mammography, and fluoroscopy.

Table 8.1 Parameters for Digital X-Ray Imaging Systems [1]

Clinical Task→	Chest Radiology	Mammography	Fluoroscopy
Detector size	35 cm × 43 cm	18 cm × 24 cm	25 cm × 25 cm
Pixel size	200 μm × 200 μm	50 μm × 50 μm	250 μm × 250 μm
Number of pixels	1,750 × 2,150	3,600 × 4,800	1,000 × 1,000
Readout time	~1 s	~1 s	1/30 s
X-ray spectrum	120 kVp	30 kVp	70 kVp
Mean exposure	300 μR	12 mR	1 μR
Exposure range	30–300 μR	0.6–240 mR	0.1–10 μR
Radiation noise	6 μR	60 μR	0.1 μR

8.2 The Ideal X-Ray Photoconductor

The first photoconductor that was found to be highly successful was amorphous selenium, just as in electrophotography. This is based on several unique features of a-Se. For example, it can coat a thick film over large areas and still maintain its properties uniformly. Another key requirement is that it is also highly X-ray sensitive. Amorphous selenium in the form of large photoreceptor plates was originally used in xeroradiography, which is essentially the photocopying of a body part using X-rays instead of light [9]. However, this system suffers from the difficulties and noise associated with the powder development technique. Xeroradiography is no longer competitive because of the toner readout method, not the underlying properties of the a-Se photoconductor [10]. By replacing the toner readout with an electrostatic readout, a-Se has again become essential to the clinical imaging system, and commercial interest in a-Se has been revived.

The flat-panel X-ray image detectors that use an a-Se photoconductor, described in Section 8.1, provide excellent images (see Figure 8.4). Amorphous selenium may not be the only choice—just as present xerographic photoreceptors eventually moved from a-Se to new man-made (organic) photoconductors that had distinct economic advantages [11]. It is therefore instructive to identify what constitutes a nearly perfect photoconductor in order to improve their performance and find better materials. Ideally, the photoconductive layer should possess the following material properties [3]:

a. The photoconductor should have as high an intrinsic X-ray sensitivity as possible. That is, it must be able to generate as many collectible (free) EHPs as possible per unit of incident radiation.

b. Nearly all the incident X-ray radiation should be absorbed within a practical photoconductor thickness to avoid unnecessary exposure to the patient.

c. There should be no dark current. This means the contacts to the photoconductor should be noninjecting and the rate of thermal generation of carriers from various defects or states in the band gap should be negligibly small. In the case of electrostatic photoconductors, the dark discharge of the surface potential (in the absence of X-rays) should be negligible.

d. There should be no bulk recombination of electrons and holes as they drift to the collection electrode; EHPs are generated in the bulk of the photoconductor.
e. There should be no deep trapping of EHPs which means that, for both electrons and holes, the schubweg is $\mu\tau E >> L$, where μ is the drift mobility, τ is the deep trapping time (lifetime), E is the electric field, and L is the layer thickness.
f. The longest carrier transit time, which depends on the smallest drift mobility, must be shorter than the access time of the pixel and inter-frame time in fluoroscopy. The transit time in the electrostatic readout is not constant because the surface voltage is photodischarged by X-ray radiation [12].
g. The transit time should not change or deteriorate with time and as a consequence of repeated exposure to X-rays. X-ray fatigue and X-ray damage should be negligible.
h. The photoconductor should be easily coated onto the AMA panel by conventional vacuum techniques, for example. Special processes are generally more expensive. In the case of electrostatic readout, the photoconductor should be coatable on a large-area conducting substrate.

Typically, the large-area coating requirement in (h) is for areas 30 cm × 30 cm or greater, which rules out the use of X-ray sensitive crystalline semiconductors since they are difficult to grow in such large areas. Various polycrystalline semiconductors such as $Zn_yCd_{1-y}Te$ and PbI_2 may be prepared in large areas, but their main drawback is the adverse effect of grain boundaries limiting charge transport. Further drawbacks are the high substrate and annealing temperatures required to optimize the semiconductor properties. Organic photoconductors that currently dominate the xerographic photoreceptor industry and can be cheaply prepared in large areas are useless because they do not satisfy (a) and (b). On the other hand, amorphous semiconductors such as a-Se, a-As_2Se_3, and a-Si:H are routinely prepared in large areas for such applications as xerographic photoreceptors and solar cells and are therefore well suited for flat-panel X-ray detector application. Among the three, a-Se is particularly well poised because it has a much greater X-ray absorption coefficient than a-Si:H, due to greater Z (atomic number). It also possesses good charge transport properties for both holes and electrons compared with a-As_2Se_3, in which electrons become trapped and hole mobility is much smaller. In addition, dark current in a-Se is much smaller than that in a-As_2Se_3. Because of its commercial use as an electrophotographic photoreceptor, a-Se is one of the most highly developed photoconductors [13]. It can be easily coated as thick films (e.g., 100–150 μm) onto suitable substrates by conventional vacuum-deposition techniques and without the need of a substrate temperature beyond 60–70°C. Its amorphous state maintains uniform characteristics for very fine scales over large areas. A large-area detector (e.g., at least 24 × 18 cm for mammography) is esential in radiography since the lack of a practical means to focus X-rays necessitates a shadow X-ray image that is larger than the imaged body part.

8.3 Intrinsic Resolution of X-Ray Photoconductors

Photoconductors that directly convert the X-ray radiation to EHPs have a number of distinct advantages, one of which is their intrinsic high resolution. The resolution of an imaging device is specified in terms of its modulation transfer function (MTF), which is the relative response of the system as a function of spatial frequencies. The higher the MTF, the better the resolution can be. It is instructive to examine the intrinsic resolution of a photoconductor-based detector. Consider an electroded a-Se layer that has been biased to establish field F in the photoconductor, and assume that the pixel size is negligible. X-rays absorbed in the photoconductor release EHPs. Holes are drawn to the top electrode and become neutralized; electrons accumulate on the storage capacitance and form the latent charge image.

The lateral spreading of information and, hence, the loss of resolution in a photoconductor-based detection system can be attributed to a number of causes. The extension of the ideas of Que and Rowlands [14] to the present electroded system suggests the following causes for the loss of resolution [1]:

1. The range of primary electrons generated by photoelectric effect.
2. Reabsorption of characteristic K-fluorescent X-rays away from the original photoelectric absorption site.
3. Reabsorption of Compton scattered photons.
4. Internal diffusion of drifting X-ray-photogenerated charge carriers as they traverse the photoconductor thickness.
5. Internal spreading due to the internal field arising from injected carriers, that is space-charge effects that arise as a result of the charge of the injected carriers or Coulombic repulsion between the drifting charge of the same sign.
6. Induced charges in neighboring pixels due to trapped charges in the photoconductor.
7. Bulk space-charge due to trapped carriers perturbing the field, which modifies the photogeneration process and changes the charge carrier transport and collection characteristics.
8. Geometric blurring due to the oblique incidence of X-rays and finite-photoconductor thickness.

These mechanisms are schematically illustrated in Figure 8.5. Que and Rowlands found that the range of primary electrons generated by the absorbed X-ray photons and oblique X-ray incidence effects limit the resolution of the a-Se photoconductor. They were able to conclude that the inherent resolution of the a-Se photoconductor system is far superior to that of the CsI-based columnar phosphor system. The range of primary electrons that is generated by an absorbed photon depends on its energy and the density of the material. This range is typically ~1–3 μm at 10–30 keV and ~10–30 μm at 50–100 keV. K-fluorescent X-rays may be released after the interaction of an X-ray photon with the K-shell of an atom. The fluorescent X-rays are released isotropically and can be reabsorbed at a point distant from their creation, thus giving rise to a characteristic type of blurring above the K edge of selenium. Geometric blurring arises when X-rays are of oblique incidence. Since photons are absorbed at different depths, they give a different response at the collection pixels, depending on the depth of absorption. For an a-Se photoconductor of thickness 200–1,000 μm and for the largest angle of incidence of the order of 15°, the blurring can be of 50–250 μm (a significant amount compared to pixel size).

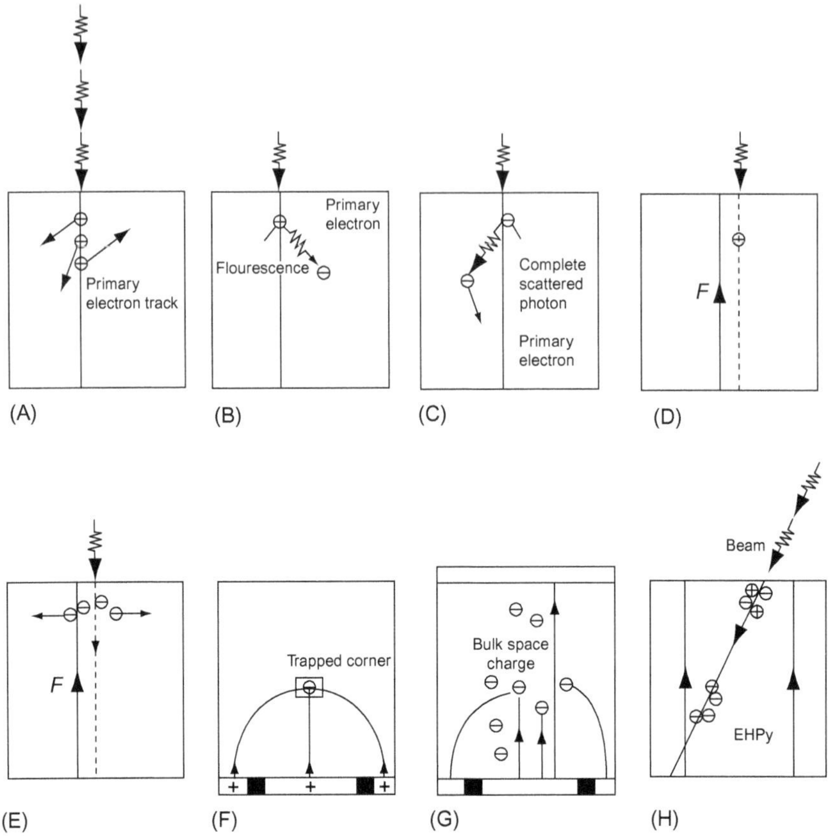

Figure 8.5 Various mechanisms that can lead to the broadening of the image and hence a loss of resolution in a photoconductor-based X-ray imaging system. (A) Lateral spreading due to the range of the primary electrons, (B) lateral spreading due to the reabsorption of K-fluorescence, (C) lateral spreading due to resorption of a Compton scattered photon, (D) lateral carrier diffusion during drift, (E) lateral spreading due to space-charge effects (Coulombic repulsion), (F) lateral spreading due to bulk trapped carriers (the average filed is relatively unaffected), (G) lateral spreading due to bulk space charge modifying the field, and (H) lateral spreading due to oblique incidence [1].

8.4 Absorption, Photoconductor Thickness, and Carrier Schubwegs

The photoelectric absorption coefficient α represents the interaction of the incident X-ray photon with the atoms in the material to produce an energetic primary electron. This electron then goes on to create many EHPs. If Φ is the fraction of incident X-ray photons that are absorbed by the photoconductor, then $\Phi = 1 - \exp[-\alpha(E, Z, d)L]$, where L is the detector thickness. $\alpha(E, Z, d)$ is the absorption coefficient of

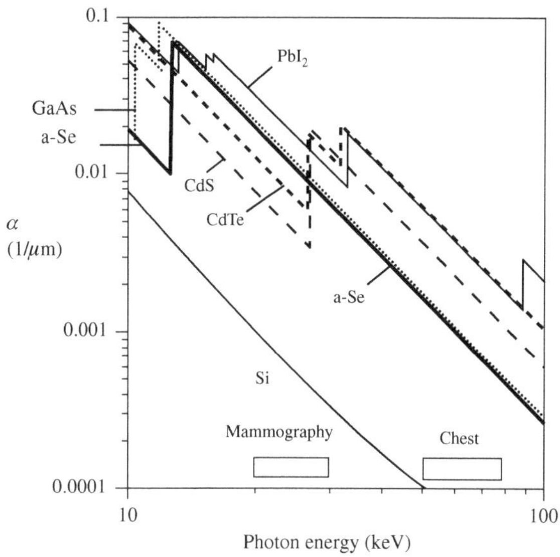

Figure 8.6 Absorption coefficient (1/μm) versus photon energy (keV) [1,3].

the photoconductor material, which depends on the energy E of the incident X-ray photons and the atomic number Z, and density d of the material. Figure 8.6 shows the energy dependence of α for a selection of photoconductors. The initial interaction of an X-ray photon with an atom of the material leads to the emission of an energetic electron from an inner core, such as the K-shell, into the conduction band. This is the photoelectric effect, and the α versus photon energy curves in Figure 8.6 correspond to the sharp vertical edges. Each shell marks an onset of absorption and as the energy increases, the absorption coefficient decreases, typically as E^{-n} where $n \approx 3$, for example. The absorption coefficient increases with atomic number Z of the material, typically as $\alpha\, Z^n$, where $n \approx 3$–5. The low Z is the primary reason for inexpensive organic semiconductors and a-Si:H being excluded as X-ray photoconductor candidates.

The minimization of dosage requires α to be such that most of the radiation is absorbed within L, or $1/\alpha < L$. The required photoconductor thickness has to be several times the absorption depth $1/\alpha$, which means that it depends on the photon energy, hence the particular imaging application and the location of the K and L edges. For mammography with 20 keV, the required a-Se thickness, taken as $2/\alpha$, is about 100 μm, whereas it is about 2,000 μm for chest radiology with a mean photon energy of 60 keV. In comparison, the corresponding thicknesses for a CdTe detector are 160 and 500 μm, respectively. The distinct advantage of a-Se is that it can be conveniently prepared as uniform thick layers over large areas by simple vacuum deposition without harming the underlying AMA substrate. Growth of single crystals of GaAs, CdTe, CdZnTe, ThBr, and other potential X-ray photoconductors is technologically limited to small areas and not suitable for large-area detection; typically, the diameters do not exceed a few inches. Although it is potentially possible to prepare large-area polycrystalline layers of PbO and PbI_2, the former has stability problems and the latter has limited sensitivity due to charge carrier schubweg limitation.

An important parameter that controls the charge collection efficiency η of X-ray-generated charge carriers is the schubweg per unit thickness, $s = \mu\tau F/L$. The actual collection efficiency depends not only on the electron and hole schubwegs but also on the attenuation coefficient of the material that determines the distribution of photogenerated carriers. Nonetheless, $s >> 1$ for both electrons and holes is still a necessary condition for near perfect collection: $\eta \approx 1$. In addition, whether a hole or electron schubweg is the primary controlling factor in the collection efficiency depends on the polarity applied to the radiation receiving (front) electrode and the absorption depth per unit thickness (δ/L). There is an advantage to applying a negative polarity to the front electrode since it avoids damage to the TFT switch on exposure. Positive polarity requires more complicated protection schemes.

8.5 Medical Applications

The medical applications for which flat-panel detectors are being developed and the new opportunities they allow will be discussed in the context of current clinical practice. All clinical applications will benefit from the following general features of flat-panel detectors:

- Immediate readout after radiation exposure to verify patient position and appropriate image exposure.
- Digital storage and communication within the hospital and beyond.
- Facilitation of computer-aided diagnosis and second opinions.
- The possibilty of improving image quality without increasing patient X-ray exposure due to enhanced detective quantum efficiency.

Another less recognized benefit of the flat-panel detector is its inherent computer control—the majority of modern X-ray machines are microprocessor controlled. In addition to computer control of X-ray energy, computer synchronization is possible for the delivery of X-ray exposure, acquisition and readout of the image, and movement of the X-ray tube and other mechanical devices such as filter holders. Complex imaging procedures, such as dual energy and tomographic data acquisition, are computer controlled to improve the busy clinical environment. The following list describes the requirements of important clinical imaging tasks and the improvements made possible by digital X-ray imaging and flat-panel X-ray detectors:

a. *Chest radiography*: Flat-panel active-matrix X-ray imagers have been configured for chest imaging. The foremost requirements are a very large field of view, a reasonably high spatial resolution (100–200 μm pixels), and a very large dynamic range to accommodate the different penetration of the lungs and mediastinum. Digital image processing can be used to equalize the appearance of the image and thus a lower X-ray beam will be used in the future.

b. *Mammography*: Mammography is the only projection X-ray imaging modality that attempts to visualize soft tissue contrast. Thus, it requires very highly absorbing beams. Film/screen is the current gold standard, but it has a small dynamic range. Therefore, extreme breast compression is needed to equalize the X-ray path length so that the whole breast can be visualized. Digital mammography is still undergoing development, but it has the potential advantages of increased dynamic range, less breast compression, and the capacity to visualize denser breasts. The challenge is to make pixels that are small enough at an affordable cost. The intrinsically high resolution of a-Se combined with the relative simplicity of the AMA design used for direct conversion suggests that this may be an ideal approach for mammography.

c. *Fluoroscopy*: Perhaps the most demanding potential application for flat-panel imaging systems is fluoroscopy. Very high patient doses result from lengthy interventional fluoroscopic procedures. During these procedures, low radiation exposure rates must be used to reduce the total exposure to the patient. This sets a stringent limit on the system performance, since the image quality must be adequate for visualization of the interventional tools as well as the anatomy. Therefore, the imaging system must be X-ray quantum limited, even at extremely low exposure levels, which implies a very low system noise for the AMA. The current technology uses a large vacuum tube device—an X-ray intensifier. Active-matrix panels are more compact, permit better access to patients, and since the panel is flat, it is largely free from geometrical distortions characteristic of vacuum tube image intensifiers. Thus, quantitative image analysis, registration, and clinical comparison of images from other modalities, three-dimensional construction (e.g., cone beam volume computerized tomography), and use in conjunction with magnetic-resonance imaging are facilitated [1].

8.6 Glance to the Future

Flat-panel X-ray detectors are still relatively new. Many advances in system design and performance can be expected. As fabrication techniques and device yields improve, more sophisticated switching structures with reduced coupling capacitance, lower leakage currents, smaller physical area, and more robust operating characteristics will continue to be developed. These advances will improve the imaging performance of AMAs until the dominant factors become the properties of the X-ray detection medium, even for the most demanding low signal level and high-resolution applications, such as fluoroscopy and mammography.

The investigation of large-area flat-panel sensors presents a large variety of previously unexplored problems in detector physics. How they may be resolved has been discussed. As technology develops, an essentially ideal X-ray imaging detector will be possible. Though the initial investment has been high, the image quality and labor saving will justify this investment over time. Eventually, mass production will reduce the costs.

8.7 References

1. S.O. Kasap, J.A. Rowlands, Proc. IEEE, 90 (2002) 391.
2. S.O. Kasap, J.B. Frey, G. Belev, O. Tousignant, H. Mani, L. Laperriere, A. Reznik, J.A. Rowlands, Phys. Status Solidi B 246 (2009) 1794.
3. S.O. Kasap, J.A. Rowlands, J. Mater. Sci. Mater. Electron. 11 (2000) 179.
4. S.O. Kasap, J. Rowlands, B. Fogal, M.Z. Kabir, G. Belev, N. Sidhu, B. Polishchuk, R.E. Johanson, J. Non-Cryst. Solids 299–302 (2002) 988.

5. D.L. Lee, I.K. Cheung, I. Jeromin, Proc. SPIE 2432 (1995) 237.
6. D.L. Lee, I.K. Cheung, E.F. Palecki, I. Jeromin, Proc. SPIE 2708 (1996) 511.
7. US Philips Corporation, USA, 1995, US Patent 5 396 072 (7 May 1995) X-ray Image Detector.
8. K. Suzuki, in: Amorphous and Microcrystalline Semiconductor Devices: Optoelectronic Devices (ed. I. Kanicki) (Artech House, Boston, MA, 1991) Chapter 3.
9. J.W. Boag, Phys. Med. Biol. 18 (1973) 3.
10. I. Brodie, R.A. Gutcheck, Med. Phys. 12 (1985) 362.
11. D.M. Pai, B.E. Springett, Rev. Mod. Phys. 65 (1993) 163.
12. S.O. Kasap, V. Aiyah, B. Polischuk, A. Baillie, J. Appl. Phys. 83 (1998) 2879.
13. S.O. Kasap, in: Handbook of Imaging Materials (eds. A.S. Diamond and D.S. Weiss) (Marcel Dekker, Inc., New York, 2nd ed., 2002) p. 329, and references therein.
14. W. Que, J.A. Rowlands, Med. Phys. 22 (1995) 2029.

www.ingramcontent.com/pod-product-compliance
Ingram Content Group UK Ltd.
Pitfield, Milton Keynes, MK11 3LW, UK
UKHW020124290926
15471UKWH00003B/270

9 780323 165396